Student Supplement
for

Intermediate
Algebra

Student Supplement for
Intermediate Algebra

Daniel L. Auvil
Kent State University

Addison-Wesley Publishing Company
Reading, Massachusetts • Menlo Park, California
London • Amsterdam • Don Mills, Ontario • Sydney

Reproduced by Addison-Wesley from camera-ready copy prepared by the author.

ISBN 0-201-00136-5
  DEFGHIJK-AL-8987654321

Preface

This supplement is designed to aid those students who are studying mathematics from the text <u>Intermediate Algebra</u> by Daniel L. Auvil. It contains step-by-step solutions to all of the even-numbered problems in the text.

Besides providing detailed solutions to the even-numbered problems, the supplement can also be helpful in solving the odd-numbered problems. Since the problems in the text are generally matched so that Problem 1 is similar to Problem 2, Problem 3 is similar to Problem 4, and so on, the solution to a particular odd-numbered problem can usually be determined by noting the procedure for solving the corresponding even-numbered problem.

Instructors can also use the supplement to their advantage by assigning even-numbered problems for homework. This reduces the amount of class time spent answering questions on homework problems, and it allows more time for introducing new topics in a more thorough and enjoyable fashion.

North Canton, Ohio   D.L.A.
January 1979

CONTENTS

CHAPTER ONE

THE REAL NUMBER SYSTEM

Section 1.1

2. T   4. T   6. F   8. F

10. T   12. Finite   14. Infinite   16. Infinite

18. Finite   20. Infinite   22. {6, 7, 8, 9} 24. ϕ

26. {T, a, ℓ, h, s, e}   28. {101, 102, 103, ···}

30. {100, 101, 102, ···, 999}

32. {x|x is a natural number between 6 and 12}

34. {x|x is a natural number greater than 30}

36. {x|x is a natural number between 1 and 2}

38. {x|x is an even natural number}

40. {x|x is a continent}

Section 1.2

2.T   4. T   6. T   8. F

10. T   12. T   14. T   16. F

18. ϕ, {6}   20. ϕ, {3}, {4}, {3, 4}

22. ϕ, {a}, {b}, {c}, {a, b}, {a, c}, {b, c}, {a, b, c}

24. A ⊆ B means every element of A is also an element of B. B ⊆ A

    means every element of B is also an element of A. Therefore,

    A and B have exactly the same elements. That is, A = B.

26. 10 ways   28. {3, 5}

30. {1, 2, 3, 4, 5, 7}   32. ϕ

34. A

36. A ∩ {2, 3, 5, 6, 7, 8} = {2}

38. {2} ∪ C = {2, 3, 5, 6, 8}

40. ∅ ∪ B = B

42. ∅

44. H, since S ⊆ H

46. S, since S ⊆ T

48. {2}

50. ∅

52. m + n - 2

Section 1.3

2. (a) 1, 119 (b) 1, 119, -0 (c) 1, 119, -0, -100 (d) all of them

4. 0.25   6. 7.8   8. -0.875   10. $0.\overline{4}$

12. $0.\overline{18}$ 14. $0.5\overline{90}$

16. 0.17, $0.\overline{17}$, 0.177, $0.1\overline{7}$ 18. 1.08, $1.\overline{08}$, $\frac{9}{5}$, $1.\overline{8}$

20. $\frac{2}{3}$

22. Rational: $\frac{1}{4}$, 0, -51, $0.05\overline{05}$, $\sqrt{36}$, $\frac{22}{7}$

   Irrational: $\sqrt{7}$, -π, 0.8181181118 ⋯, e

24. T   26. T   28. F   30. T

Section 1.4

2.

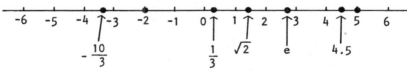

4. $|14| = 14$, $|-8| = 8$, $|-4\frac{2}{3}| = 4\frac{2}{3}$, $|-\sqrt{3}| = \sqrt{3}$, $|+0| = 0$

6. $-8 > -10$ 8. $0 > -1$ 10. $-7 < 4$ 12. $|-7| > 4$

14. $π > 3.14$ 16. $|-4| - |4| = 4 - 4 = 0$

18. $|-6| + |-3| = 6 + 3 = 9$ 20. $-|-15| = -15$

22. $x < 0$ 24. $x \leq 0$ 26. $x > 7$ 28. $x - 5 \leq 6$

30. $-2 < x < 0$

32. 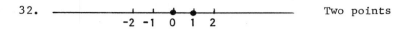 Two points

   -2 -1 0 1 2

34. Infinite number of points

   -2 -1 0 1 2

36. Open half-line

   -2 -1 0 1 2

38. Closed half-line

   -2 -1 0 1 2

40. Open half-line

   -2 -1 0 1 2

42. Closed interval

   -2 -1 0 1 2

44. Half-open interval

   -2 -1 0 1 2 3 4

46. Closed interval

   -2 -1 0 1 2 3 4

48. No graph

   -2 -1 0 1 2

50. Line

   -2 -1 0 1 2

52. $2.9 \text{ cm} \leq D \leq 3.1 \text{ cm}$ 54. $0 \text{ min} \leq \ell \leq 6 \text{ min}$

Section 1.5

2. Inverse for addition   4. Identity for multiplication

6. Identity for addition   8. Distributive

10. Inverse for multiplication

12. Commutative for multiplication

14. Associative for addition

16. Double negative theorem

18. Multiplication theorem of zero

20. Multiplication theorem of equality

22. No. $4 \div 2 \neq 2 \div 4$

24. No. $3 - (2 - 1) \neq (3 - 2) - 1$

26. 7   28. -1   30. Positive   32. $-(1 - \sqrt{3})$

34. $-(a - b)$

Section 1.6

2. -7   4. 4   6. -2   8. -20

10. 17   12. 9   14. -15   16. 2

18. -2.5   20. 12   22. 5   24. 0

26. $3 - 5 = 3 + (-5) = -2$ 28. $5 - 10 = 5 + (-10) = -5$

30. $(-2) - 7 = (-2) + (-7) = -9$ 32. $(-8) - 5 = (-8) + (-5) = -13$

34. $6 - (-4) = 6 + 4 = 10$ 36. $7 - (-13) = 7 + 13 = 20$

38. $0 - 8 = 0 + (-8) = -8$ 40. $(-9) - (-1) = (-9) + 1 = -8$

42. $(-14) - (-16) = (-14) + 16 = 2$

44. $0 - (-2) = 0 + 2 = 2$

46. $(-8.7) - 5.8 = (-8.7) + (-5.8) = -14.5$

48. $4.9 - 9.2 = 4.9 + (-9.2) = -4.3$

50. $\dfrac{3 + 1 + 0 + (-4) + (-2) + 4 + 5}{7} = \dfrac{7}{7} = 1°$

52. $10 - (3 \cdot 4) = 10 - 12 = -2°$

54. $T = 10 - 3h$

    a) $T = 10 - (3 \cdot 4) = 10 - 12 = -2°$

    b) $T = 10 - (3 \cdot 5) = 10 - 15 = -3°$

    c) $T = 10 - (3 \cdot 10) = 10 - 30 = -20°$

Section 1.7

2. -20	4. 16	6. -42	8. -19
10. 0	12. -200	14. -60	16. -234
18. 6.25	20. 891	22. 16	24. 64
26. -3	28. 5	30. -4	32. -4
34. -1	36. 6	38. 0	40. 12

42. -7.3

44. $-\dfrac{16}{8} = -2$ but $\dfrac{-16}{-8} = 2$

46. $-\dfrac{-3}{5} = \dfrac{-(-3)}{5} = \dfrac{3}{5}$

48. $-\dfrac{1}{-2} = \dfrac{-1}{-2} = \dfrac{1}{2}$

50. $-\dfrac{-x}{20} = \dfrac{-x}{-20} = \dfrac{x}{20}$

52. total yardage $= 3 + (-12) + (-6) + 7 = -8$ yd

    average gain $= \dfrac{-8}{4} = -2$ yd

54. $T = 10 + 3h$

    a) $T = 10 + (3 \cdot 10) = 10 + 30 = 40°$

    b) $T = 10 + [3(-3)] = 10 + [-9] = 1°$

    c) $T = 10 + [3(-4)] = 10 + [-12] = -2°$

Section 1.8

2. $5 + 3 \cdot 2$

    $= 5 + 6$

    $= 11$

4. $10 - 4 \cdot 2$

    $= 10 - 8$

    $= 2$

6. $-6 + \dfrac{12}{3}$

   $= -6 + 4$

   $= -2$

8. $(5 + 9)2$

   $= (14)2$

   $= 28$

╳10. $16 - 3(7 + 2)$

   $= 16 - 3(9)$

   $= 16 - 27$

   $= -9$

12. $(-8) \cdot 9 + 3 \cdot 6$

   $= -72 + 18$

   $= -54$

14. $\dfrac{-16}{8} + \dfrac{-12}{-4}$

   $= -2 + 3$

   $= 1$

16. $\dfrac{7 + (-3)}{2} - 2$

   $= \dfrac{4}{2} - 2$

   $= 2 - 2$

   $= 0$

18. $\dfrac{-100}{6 - (-4)}$

   $= \dfrac{-100}{6 + 4}$

   $= \dfrac{-100}{10}$

   $= -10$

20. $4[7 - (9 + 10)]$

   $= 4[7 - 19]$

   $= 4[-12]$

   $= -48$

22. $3(2[12 - 4(8 - 6)])$

   $= 3(2[12 - 4(2)])$

   $= 3(2[12 - 8])$

   $= 3(2[4])$

   $= 3(8)$

   $= 24$

24. $8[-13 + 2(5 - 13)]$

   $= 8[-13 + 2(-8)]$

   $= 8[-13 + (-16)]$

   $= 8[-29]$

   $= -232$

26. $21 \div [-3 - 5(7 - 9)]$

   $= 21 \div [-3 - 5(-2)]$

   $= 21 \div [-3 - (-10)]$

   $= 21 \div [-3 + 10]$

   $= 21 \div [7]$

   $= 3$

28. $\dfrac{4(-1) + 3(-2)}{-1 - 1}$

   $= \dfrac{-4 + (-6)}{-1 + (-1)}$

   $= \dfrac{-10}{-2}$

   $= 5$

30. $\left[\dfrac{(-7) + 3}{-1 + 5}\right]\left[\dfrac{28 + (-10)}{2 - 8}\right]$ 32. $\dfrac{(-5)4 - (-8)}{(-4) + 1} - (-2)^2$

$= \left[\dfrac{-4}{4}\right]\left[\dfrac{18}{-6}\right]$ $= \dfrac{-20 + 8}{-3} - 4$

$= [-1]\,[-3]$ $= \dfrac{-12}{-3} - 4$

$= 3$ $= 4 - 4$

$= 0$

CHAPTER TWO
POLYNOMIALS

Section 2.1

2. $5^3 = 5 \cdot 5 \cdot 5 = 125$

4. $(-1)^5 = (-1)(-1)(-1)(-1)(-1) = -1$

6. $2 \cdot 4^2 = 2 \cdot 16 = 32$

8. $(-9)^2 = (-9)(-9) = 81$ and $-9^2 = -(9 \cdot 9) = -81$.

10. $(-3)^3 = (-3)(-3)(-3) = -27$ and $-3^3 = -(3 \cdot 3 \cdot 3) = -27$.

12. 7^6

14. $36x^2y^4$

16. $(3y + 1)^3$

18. $a^2b^2 - 13a^2b^3$

20. $(2x)^3 + ab^2$

22. $x^4 \cdot x^5 = x^{4+5} = x^9$

24. $a^{19} \cdot a \cdot a^{38} = a^{19+1+38} = a^{58}$

26. $9^3 \cdot 9^9 = 9^{3+9} = 9^{12}$

28. $(x - 4)^7 (x - 4)^7 = (x - 4)^{7+7} = (x - 4)^{14}$

30. $\dfrac{x^{20}}{x^{10}} = x^{20-10} = x^{10}$

32. $\dfrac{a^4 b^{20}}{ab^5} = a^{4-1} b^{20-5} = a^3 b^{15}$

34. $\dfrac{7^9}{7^3} = 7^{9-3} = 7^6$

36. $\dfrac{(y + 3)^{12}}{(y + 3)^4} = (y + 3)^{12-4} = (y + 3)^8$

38. $(x^2)^3 = x^{2 \cdot 3} = x^6$

40. $(b^{10})^7 = b^{10 \cdot 7} = b^{70}$

42. $(ab)^4 = a^4b^4$

44. $(5xy)^3 = 125x^3y^3$

46. $(\dfrac{x}{y})^5 = \dfrac{x^5}{y^5}$ 48. $(\dfrac{2}{b})^4 = \dfrac{16}{b^4}$

50. $(x^4 y^2)^3 = (x^4)^3 (y^2)^3 = x^{12} y^6$

52. $(-2x^6 y)^5 = (-2)^5 (x^6)^5 y^5 = -32x^{30} y^5$

54. $(\dfrac{5a^3}{b})^2 = \dfrac{5^2 (a^3)^2}{b^2} = \dfrac{25a^6}{b^2}$

56. $(\dfrac{2a^8}{3b^6})^4 = \dfrac{2^4 (a^8)^4}{3^4 (b^6)^4} = \dfrac{16a^{32}}{81b^{24}}$

58. $(a + b)^5 (a + b)^5 = (a + b)^{5+5} = (a + b)^{10}$

60. $3(ab^2)^3 = 3a^3 (b^2)^3 = 3a^3 b^6$

62. $x^3 (x^2 y)^6 = x^3 (x^2)^6 y^6 = x^3 x^{12} y^6 = x^{15} y^6$

64. $\dfrac{-75(a^5 b)^2}{25a^2 b^2} = \dfrac{-75a^{10} b^2}{25a^2 b^2} = -3a^8$

66. $(y^n)^4 = y^{4n}$ 68. $x^n \div x = x^n \div x^1 = x^{n-1}$

70. After 1 hr, the culture contained $45 = 15 \cdot 3^1$ cells.

   After 2 hr, the culture contained $135 = 15 \cdot 3^2$ cells.
   $\vdots$
   After 24 hr, the culture contained $15 \cdot 3^{24}$ cells.

72. $2 = 2^1$ ancestors were part of the last generation.

   $4 = 2^2$ ancestors were part of the generation 2 generations ago.
   $\vdots$
   2^{60} ancestors were part of the generation 60 generations ago.

74. The number of cells after h hours is $5 \cdot 4^h$.

   If this is to equal 320, we have

   $$5 \cdot 4^h = 320$$
   $$4^h = 64$$
   $$4^h = 4^3$$
   $$h = 3 \text{ hr.}$$

Section 2.2

2. a, b, f

4.
	Degree	Leading coefficient	Constant term
a)	3	15	7
b)	2	1	$\sqrt{3}$
c)	5	1	0
d)	88	5	10
e)	3	6	$\frac{2}{3}$
f)	7	-1	1

6. a) 1st degree binomial

b) 5th degree monomial

c) 3rd degree binomial

d) 2nd degree trinomial

e) 7th degree trinomial

f) 3rd degree monomial

8. $P(2) = 2^2 - 7 \cdot 2 + 1 = 4 - 14 + 1 = -9$

10. $P(0) = 0^5 - 0^3 + 0 = 0$

12. $Q(-3) = 2(-3)^2 - 8(-3) - 9 = 2 \cdot 9 + 24 - 9 = 18 + 24 - 9 = 33$

14. $R(\frac{2}{3}) = 3(\frac{2}{3})^3 + (\frac{2}{3})^2 + (\frac{2}{3}) + \frac{1}{9}$

$$= 3(\frac{8}{27}) + \frac{4}{9} + \frac{2}{3} + \frac{1}{9}$$

$$= \frac{8}{9} + \frac{4}{9} + \frac{4}{9} + \frac{1}{9}$$

$$= \frac{17}{9}$$

16. $P(-5) = -(-5)^2 + 10 = -25 + 10 = -15$

18. $H(\sqrt{2}) = (\sqrt{2})^4 + 1 = 4 + 1 = 5$

20. $Q(-1) = 3(-1) - 2 = -5$ and $P(-5) = (-5)^2 = 25.$

Therefore, $P(Q(-1)) = P(-5) = 25.$

22. $P(0) = 0^3 + 5 = 5$ and $Q(4) = -4^2 = -16$.

Therefore, $P(0) + Q(4) = 5 + (-16) = -11$.

24. $P(x) = ax^2 + bx + c$, where a, b, and c are constants.

26. $P(x) = a_n x^n + a_{n-1} x^{n-1} + \cdots + a_2 x^2 + a_1 x + a_0$, where

$a_n, a_{n-1}, \cdots, a_2, a_1,$ and a_0 are constants.

28. $h(t) = -16t^2 + 32t + 48$

a) $h(1) = -16(1)^2 + 32(1) + 48 = 64$ ft

b) $h(2) = -16(2)^2 + 32(2) + 48 = 48$ ft

c) $h(3) = -16(3)^2 + 32(3) + 48 = 0$ ft

d) $h(4) = -16(4)^2 + 32(4) + 48 = -80$

The formula does not apply in this last case.

30. $C(75) = 3(75) + 50 = 225 + 50 = \$275$

Section 2.3

2. $9x$

4. $11z^2$

6. 0

8. $-8st$

10. $-10y + 5$

12. $7m^2 + 9m$

14. $2a^3$

16. $-5ab^3 - a^3b$

18. $-7r$

20. $13x + 11$

22. $10x^2 + 8x + 3$

24. $9y^3 - 6y^2 - 17$

26. $2a^3 - 6a^2 + 5a + 6$

28. $10x^3y - 6x^2y + 5xy - 2y^2$

30. $6x^2 + 2x + 5$

32. $-3y^2 + 15y + 2$

34. $2a^3 + 6a^2 - 4a + 4$

36. $x^2 - 14y^2$

38. $3(x^3 - x^2 + 4x + 6) + (3x^3 + 7x^2 + 1)$

$= 3x^3 - 3x^2 + 12x + 18 + 3x^3 + 7x^2 + 1$

$= 6x^3 + 4x^2 + 12x + 19$

40. $5(x^3 + 2x^2 - x) - 2(-4x^2 + x - 2)$

$= 5x^3 + 10x^2 - 5x + 8x^2 - 2x + 4$

$= 5x^3 + 18x^2 - 7x + 4$

42. $7(4x^3 - 3xy + y^2) - (17x^3 - xy - y^2)$

   $= 28x^3 - 21xy + 7y^2 - 17x^3 + xy + y^2$

   $= 11x^3 - 20xy + 8y^2$

44. $(a - 2b + c) + (2a - b + c) - (a + b + 2c)$

   $= a - 2b + c + 2a - b + c - a - b - 2c$

   $= 2a - 4b$

46. $3 + 2(x - 10) + (5x + 6)$ 48. $2 + 9[(x + 5) - (x - 7)]$

   $= 3 + 2x - 20 + 5x + 6$ $= 2 + 9[x + 5 - x + 7]$

   $= 7x - 11$ $= 2 + 9[12]$

                                $= 2 + 108$

                                $= 110$

50. $2(7x + 3) + 2(3x - 4)$

   $= 14x + 6 + 6x - 8$

   $= 20x - 2$

Section 2.4

2. $(3x)(4x^2) = (3 \cdot 4)(x \cdot x^2) = 12x^3$

4. $(-9y^2)(2y^3) = (-9) \cdot 2 \cdot y^2 \cdot y^3 = -18y^5$

6. $(5ab^3)(-7a^2b^2) = 5 \cdot (-7) \cdot a \cdot a^2 \cdot b^3 \cdot b^2 = -35a^3b^5$

8. $(3xy^4)^3 (-x^2y)(2x^2y^3)$ 10. $(6r^2st)^2(6rs^2t^3)(-rt)^5$

   $= 27x^3y^{12} (-x^2y)(2x^2y^3)$ $= 36r^4s^2t^2(6rs^2t^3)(-r^5t^5)$

   $= -54x^7y^{16}$ $= -216r^{10}s^4t^{10}$

12. $x(x + 5)$ 14. $3x^2(6x - 1)$

   $= x \cdot x + x \cdot 5$ $= 3x^2 \cdot 6x - 3x^2 \cdot 1$

   $= x^2 + 5x$ $= 18x^3 - 3x^2$

16. $-7x^3(2x + 1)$

   $= (-7x^3) \cdot 2x + (-7x^3) \cdot 1$

   $= -14x^4 + (-7x^3)$

   $= -14x^4 - 7x^3$

18. $-9xy^2(x^2 - x - 7)$

$\quad = (-9xy^2)x^2 - (-9xy^2)x - (-9xy^2)7$

$\quad = -9x^3y^2 + 9x^2y^2 + 63xy^2$

20. $xy^3(x^2 - 8xy + 5y^2)$

$\quad = xy^3 \cdot x^2 - xy^3 \cdot 8xy + xy^3 \cdot 5y^2$

$\quad = x^3y^3 - 8x^2y^4 + 5xy^5$

22. $(x + 5)(x + 6)$

$\quad = x^2 + 6x + 5x + 30$

$\quad = x^2 + 11x + 30$

24. $(x - 3)(x + 8)$

$\quad = x^2 + 8x - 3x - 24$

$\quad = x^2 + 5x - 24$

26. $(2x + 7)(6x - 1)$

$\quad = 12x^2 - 2x + 42x - 7$

$\quad = 12x^2 + 40x - 7$

28. $(x + 5)(x + 5)$

$\quad = x^2 + 5x + 5x + 25$

$\quad = x^2 + 10x + 25$

30. $(2x - 5y)(2x - 5y)$

$\quad = 4x^2 - 10xy - 10xy + 25y^2$

$\quad = 4x^2 - 20xy + 25y^2$

32. $(x - 3)(x^2 + 8x - 2y)$

$\quad = x^3 + 8x^2 - 2xy - 3x^2 - 24x + 6y$

$\quad = x^3 + 5x^2 - 24x - 2xy + 6y$

34. $(a - 1)(a^3 + a^2 - 2a + 1)$

$\quad = a^4 + a^3 - 2a^2 + a - a^3 - a^2 + 2a - 1$

$\quad = a^4 - 3a^2 + 3a - 1$

36. $(a - 1)(a + 3)(a + 2)$

$\quad = (a^2 + 3a - a - 3)(a + 2)$

$\quad = (a^2 + 2a - 3)(a + 2)$

$\quad = a^3 + 2a^2 + 2a^2 + 4a - 3a - 6$

$\quad = a^3 + 4a^2 + a - 6$

38. $(7a + b)(7a - b)(a - b)$

$= (49a^2 - 7ab + 7ab - b^2)(a - b)$

$= (49a^2 - b^2)(a - b)$

$= 49a^3 - 49a^2b - ab^2 + b^3$

40. $(4c + 3)(5c^3 - 4c^2 + c - 5)$

$= 20c^4 - 16c^3 + 4c^2 - 20c + 15c^3 - 12c^2 + 3c - 15$

$= 20c^4 - c^3 - 8c^2 - 17c - 15$

42. $(7x^3y)(2xy^2) - (x^2y^2)(5x^2y)$

$= 14x^4y^3 - 5x^4y^3$

$= 9x^4y^3$

44. $x^2[x - (x + 8)(x - 1)]$

$= x^2[x - (x^2 - x + 8x - 8)]$

$= x^2[x - (x^2 + 7x - 8)]$

$= x^2[x - x^2 - 7x + 8]$

$= x^2[-x^2 - 6x + 8]$

$= -x^4 - 6x^3 + 8x^2$

46. $(x - a)(x - b)$

$= x^2 - bx - ax + ab$

$= x^2 - (a + b)x + ab$

48. $(x + a)(x + a)$

$= x^2 + ax + ax + a^2$

$= x^2 + 2ax + a^2$

50. $(x + a)(x^2 - ax + a^2)$

$= x^3 - ax^2 + a^2x + ax^2 - a^2x + a^3$

$= x^3 + a^3$

52. $w(w - 4)$

$= w^2 - 4w$

54. $\frac{1}{2}b(6b + 1)$

$= 3b^2 + \frac{1}{2}b$

Section 2.5

2. $3(x + 3)$

4. $5(x^2 + 1)$

6. $7x(x - 6)$

8. $3(5x - 6y)$

10. $a(a + b)$

12. $a(a^2 - a - 1)$

14. $3x(3x^6 + 4x^4 - x^2 - 2)$

16. $12xy(2xy - 3x + 4)$

18. $-a(bc + bd + cd)$

20. $10z^{50}(10z^{25} - 1)$

22. $xy^2z(x - 1 - z)$

24. $-5rs^2t^2(3r + 5t)$

26. $(x + y)(7x - y)$

28. $(x + y)(x + y) = (x + y)^2$

30. $(x^2 + 1)(1 - x)$

32. $(2x - 3y)(7a + 6b)$

34. $(x + y)[x(x + y) + y^2]$
   $= (x + y)[x^2 + xy + y^2]$

36. $x^n(x^2 - x - 1)$

38. $y^2(y^n + y^{n-2} + 1)$

40. $a^{2n}(a^{2n} + a^n + 1)$

Section 2.6

2. $(x + 1)(x + 2)$

4. $(x + 2)(x + 5)$

6. $(x + 2)^2$

8. $(x + 4)(x - 2)$

10. $(x - 6)(x + 4)$

12. $(x - 3)^2$

14. $(y - 2)(y - 6)$

16. $(y + 7)(y - 1)$

18. $(a - 3)(a + 2)$

20. $(a + 11)(a - 1)$

22. $(2x + 1)(x + 3)$

24. $(3x + 2)(x - 5)$

26. $(5y + 2)(y - 1)$

28. $(13y - 11)(y + 1)$

30. $(4b + 1)(3b - 5)$

32. $(4b + 1)(3b - 1)$

34. $(9x + y)^2$

36. $(4x - 3)(x - 2)$

38. $(9x + 4)(x - 2)$

40. $(4a + b)(a - 3b)$

42. $5a^2b + 60ab + 55b$
   $= 5b(a^2 + 12a + 11)$
   $= 5b(a + 1)(a + 11)$

44. $12x^3y^2 - 42x^2y^3 - 54xy^4$
   $= 6xy^2(2x^2 - 7xy - 9y^2)$
   $= 6xy^2(2x - 9y)(x + y)$

46. $(x^2 + 3)(x^2 + 4)$

48. $(3x^2 - 5)(2x^2 + 5)$

Section 2.7

2. $(x - 4)(x + 4)$

4. $(x - 10)(x + 10)$

6. $(6x - y)(6x + y)$

8. $(7x - 1)(7x + 1)$

10. $x^4 - 81$

$= (x^2 - 9)(x^2 + 9)$

$= (x - 3)(x + 3)(x^2 + 9)$

12. $x^3 + 27$

$= x^3 + 3^3$

$= (x + 3)(x^2 - 3x + 9)$

14. $x^3 - 1$

$= x^3 - 1^3$

$= (x - 1)(x^2 + x + 1)$

16. $x^3 + y^3$

$= (x + y)(x^2 - xy + y^2)$

18. $125x^3 - 8$

$= (5x)^3 - 2^3$

$= (5x - 2)(25x^2 + 10x + 4)$

20. $64x^6 + y^3$

$= (4x^2)^3 + y^3$

$= (4x^2 + y)(16x^4 - 4x^2y + y^2)$

22. $x^3 - 8x^2 + 2x - 16$

$= x^2(x - 8) + 2(x - 8)$

$= (x - 8)(x^2 + 2)$

24. $xy + 11y + 3x + 33$

$= y(x + 11) + 3(x + 11)$

$= (x + 11)(y + 3)$

26. $2x^2 - xy + 4xy - 2y^2$

$= x(2x - y) + 2y(2x - y)$

$= (2x - y)(x + 2y)$

28. $x^3 + 5x^2 + 2x + 10$

$= x^2(x + 5) + 2(x + 5)$

$= (x + 5)(x^2 + 2)$

30. $x^2 + 3xy^2 + 7xy + 21y^3$

$= x(x + 3y^2) + 7y(x + 3y^2)$

$= (x + 3y^2)(x + 7y)$

32. $10x^4 - 10y^4$

$= 10(x^4 - y^4)$

$= 10(x^2 - y^2)(x^2 + y^2)$

$= 10(x - y)(x + y)(x^2 + y^2)$

34. $a^4b^4 - 625$

$= (a^2b^2 - 25)(a^2b^2 + 25)$

$= (ab - 5)(ab + 5)(a^2b^2 + 25)$

36. $x^6 - y^6$

$= (x^3 - y^3)(x^3 + y^3)$

$= (x - y)(x^2 + xy + y^2)(x + y)(x^2 - xy + y^2)$

38. $13b + 13a^3b$

$= 13b(1 + a^3)$

$= 13b(1 + a)(1 - a + a^2)$

42. $(a - b)^2 - c^2$

$= [(a - b) - c][(a - b) + c]$

$= (a - b - c)(a - b + c)$

44. $(a - b)(a^2 + ab + b^2)$

$= a^3 + a^2b + ab^2 - a^2b - ab^2 - b^3$

$= a^3 - b^3$

40. $x^3y + y + x^3 + 1$

$= y(x^3 + 1) + (x^3 + 1)$

$= (x^3 + 1)(y + 1)$

$= (x + 1)(x^2 - x + 1)(y + 1)$

CHAPTER THREE
ALGEBRAIC FRACTIONS

Section 3.1

2. T, since $6 \cdot 15 = 9 \cdot 10$.

4. F, since $3 \cdot 9 \neq 4 \cdot 7$.

6. T, since $3 \cdot 3x^2 = (-9x) \cdot (-x)$.

8. T, since $-a(b^3c^2) = (b^3c)(-ac)$.

10. T, since $(x - 3)(2x + 6) = (x^2 - 9)2$.

12. $\dfrac{45}{150} = \dfrac{15 \cdot 3}{15 \cdot 10} = \dfrac{3}{10}$

14. $-\dfrac{-33}{-88} = -\dfrac{11 \cdot 3}{11 \cdot 8} = -\dfrac{3}{8}$

16. $\dfrac{-10a}{-15a} = \dfrac{5 \cdot 2 \cdot a}{5 \cdot 3 \cdot a} = \dfrac{2}{3}$

18. $\dfrac{20a^4b^2}{4a^2b^3} = \dfrac{4 \cdot 5a^2a^2b^2}{4a^2b^2b} = \dfrac{5a^2}{b}$

20. $\dfrac{3x - 6}{9x + 18} = \dfrac{3(x - 2)}{9(x + 2)} = \dfrac{x - 2}{3(x + 2)}$

22. $\dfrac{14x + 7y}{22x + 11y} = \dfrac{7(2x + y)}{11(2x + y)} = \dfrac{7}{11}$

24. $\dfrac{9x}{3x^2 - 15x} = \dfrac{3x \cdot 3}{3x(x - 5)} = \dfrac{3}{x - 5}$

26. $\dfrac{a^3b^3 - a^2b^2}{a^2b^2} = \dfrac{a^2b^2(ab - 1)}{a^2b^2} = \dfrac{ab - 1}{1} = ab - 1$

28. $\dfrac{x - 5}{x^2 - 25} = \dfrac{x - 5}{(x - 5)(x + 5)} = \dfrac{1}{x + 5}$

30. $\dfrac{6x + 18}{x^3 - 9x} = \dfrac{6(x + 3)}{x(x^2 - 9)} = \dfrac{6(x + 3)}{x(x - 3)(x + 3)} = \dfrac{6}{x(x - 3)}$

32. $\dfrac{(x + y)^2}{x^2 - y^2} = \dfrac{(x + y)(x + y)}{(x - y)(x + y)} = \dfrac{x + y}{x - y}$

34. $\dfrac{x^2 + 3x - 10}{x^2 + x - 6} = \dfrac{(x + 5)(x - 2)}{(x + 3)(x - 2)} = \dfrac{x + 5}{x + 3}$

36. $\dfrac{a^2 - 1}{6 - 5a - a^2} = \dfrac{(a - 1)(a + 1)}{-(a^2 + 5a - 6)} = -\dfrac{(a - 1)(a + 1)}{(a - 1)(a + 6)} = -\dfrac{a + 1}{a + 6}$

38. $\dfrac{x^3 + y^3}{4x^2 + 5xy + y^2} = \dfrac{(x + y)(x^2 - xy + y^2)}{(x + y)(4x + y)} = \dfrac{x^2 - xy + y^2}{4x + y}$

40. $\dfrac{x^2 - ax + xy - ay}{x + y}$

$= \dfrac{x(x - a) + y(x - a)}{x + y}$

$= \dfrac{(x - a)(x + y)}{x + y}$

$= x - a$

42. $\dfrac{4}{15} = \dfrac{12}{45}$

Building factor = 3

44. $\dfrac{5x}{4y^2} = \dfrac{15x^4 y}{12x^3 y^3}$

Building factor = $3x^3 y$

46. $\dfrac{1}{a - b} = \dfrac{a + b}{a^2 - b^2}$

Building factor = $a + b$

48. $\dfrac{x^2}{x + 4} = \dfrac{?}{x^2 + 3x - 4}$

$\dfrac{x^2}{x + 4} = \dfrac{?}{(x + 4)(x - 1)}$

Building factor = $x - 1$

$\dfrac{x^2}{x + 4} = \dfrac{x^2(x - 1)}{(x + 4)(x - 1)}$

50. $\dfrac{x - 3y}{2x - 2y} = \dfrac{?}{4x^2 + 8xy - 12y^2}$

$\dfrac{x - 3y}{2(x - y)} = \dfrac{?}{4(x - y)(x + 3y)}$

Building factor = $2(x + 3y)$

$\dfrac{x - 3y}{2(x - y)} = \dfrac{2(x - 3y)(x + 3y)}{4(x - y)(x + 3y)}$

Section 3.2

2. 15

4. 18

6. $40 = 2 \cdot 2 \cdot 2 \cdot 5$

$75 = 3 \cdot 5 \cdot 5$

$1cm = 2 \cdot 2 \cdot 2 \cdot 3 \cdot 5 \cdot 5 = 600$

8. $18 = 2 \cdot 3 \cdot 3$

$20 = 2 \cdot 2 \cdot 5$

$15 = 3 \cdot 5$

$1cm = 2 \cdot 2 \cdot 3 \cdot 3 \cdot 5 = 180$

10. $21x$

12. $a^2 b^2$

14. $30a^3 b^2$

16. $a - 4$

18. $x(x + 7)$

20. $(x - 3)(x + 3)$

22. $(x + 5)(x - 1)$

24. $x + 4 = x + 4$

$3x + 12 = 3(x + 4)$

$1cm = 3(x + 4)$

26. $x^2 - 9 = (x - 3)(x + 3)$

$x - 3 = x - 3$

$1cm = (x - 3)(x + 3)$

28. $2a - 2b = 2(a - b)$

$a^2 - b^2 = (a - b)(a + b)$

$1cm = 2(a - b)(a + b)$

30. $x^2 + 3x - 10 = (x + 5)(x - 2)$

$x^2 - x - 2 = (x - 2)(x + 1)$

$1cm = (x + 5)(x - 2)(x + 1)$

32. $x^2 - 9 = (x - 3)(x + 3)$

$x - 3 = x - 3$

$3x + 9 = 3(x + 3)$

$1cm = 3(x + 3)(x - 3)$

34. $(a - 1)^2$

Section 3.3

2. $\frac{4}{7} + \frac{2}{7} = \frac{4 + 2}{7} = \frac{6}{7}$

4. $\frac{13}{15} - \frac{8}{15} = \frac{13 - 8}{15} = \frac{7}{15}$

6. $\frac{x}{8} + \frac{5}{8} = \frac{x + 5}{8}$

8. $\frac{3x}{x - 1} - \frac{3}{x - 1} = \frac{3x - 3}{x - 1} = \frac{3(x - 1)}{x - 1} = 3$

10. $\frac{x - 1}{15x} - \frac{x + 2}{15x} = \frac{x - 1 - x - 2}{15x} = \frac{-3}{15x} = -\frac{1}{5x}$

12. $\frac{x}{abc} - \frac{y}{abc} + \frac{z}{abc} = \frac{x - y + z}{abc}$

14. $\frac{a}{a^2 - b^2} - \frac{b}{a^2 - b^2} = \frac{a - b}{a^2 - b^2} = \frac{a - b}{(a - b)(a + b)} = \frac{1}{a + b}$

16. $\frac{x}{x^2 + 2x - 3} - \frac{1}{x^2 + 2x - 3} = \frac{x - 1}{x^2 + 2x - 3} = \frac{x - 1}{(x + 3)(x - 1)} = \frac{1}{x + 3}$

18. $\frac{1}{3} + \frac{1}{5} = \frac{5}{15} + \frac{3}{15} = \frac{8}{15}$

20. $\dfrac{7}{18} - \dfrac{8}{9} = \dfrac{7}{18} - \dfrac{16}{18} = -\dfrac{9}{18} = -\dfrac{1}{2}$

22. $\dfrac{3x}{40} + \dfrac{13x}{75} = \dfrac{45x}{600} + \dfrac{104x}{600} = \dfrac{149x}{600}$

24. $\dfrac{1}{18} + \dfrac{9x}{20} - \dfrac{4x}{15} = \dfrac{10}{180} + \dfrac{81x}{180} - \dfrac{48x}{180} = \dfrac{33x + 10}{180}$

26. $\dfrac{1}{3x} + \dfrac{1}{7x} = \dfrac{7}{21x} + \dfrac{3}{21x} = \dfrac{10}{21x}$

28. $\dfrac{1}{a^2} - \dfrac{1}{b^2} + \dfrac{2}{ab} = \dfrac{b^2}{a^2 b^2} - \dfrac{a^2}{a^2 b^2} + \dfrac{2ab}{a^2 b^2} = \dfrac{b^2 + 2ab - a^2}{a^2 b^2}$

30. $\dfrac{x}{10a^3 b} - \dfrac{y}{15ab^2} = \dfrac{3bx}{30a^3 b^2} - \dfrac{2a^2 y}{30a^3 b^2} = \dfrac{3bx - 2a^2 y}{30a^3 b^2}$

32. $a + \dfrac{1}{a - 4} = \dfrac{a(a - 4)}{a - 4} + \dfrac{1}{a - 4} = \dfrac{a^2 - 4a + 1}{a - 4}$

34. $\dfrac{x}{x + 7} - \dfrac{1}{x} = \dfrac{x^2}{x(x + 7)} - \dfrac{x + 7}{x(x + 7)} = \dfrac{x^2 - x - 7}{x(x + 7)}$

36. $\dfrac{4x}{x - 3} + \dfrac{3}{x + 3} = \dfrac{4x(x + 3)}{(x - 3)(x + 3)} + \dfrac{3(x - 3)}{(x - 3)(x + 3)}$

$$= \dfrac{4x^2 + 12x + 3x - 9}{(x - 3)(x + 3)}$$

$$= \dfrac{4x^2 + 15x - 9}{(x - 3)(x + 3)}$$

38. $\dfrac{x - 1}{x + 5} - \dfrac{2}{x - 1} = \dfrac{(x - 1)(x - 1)}{(x + 5)(x - 1)} - \dfrac{2(x + 5)}{(x + 5)(x - 1)}$

$$= \dfrac{x^2 - 2x + 1 - 2x - 10}{(x + 5)(x - 1)}$$

$$= \dfrac{x^2 - 4x - 9}{(x + 5)(x - 1)}$$

40. $\dfrac{x}{x + 4} - \dfrac{7}{3x + 12} = \dfrac{3x}{3(x + 4)} - \dfrac{7}{3(x + 4)}$

$$= \dfrac{3x - 7}{3(x + 4)}$$

42. $\dfrac{x^2}{x^2 - 9} + \dfrac{1}{x - 3} = \dfrac{x^2}{(x - 3)(x + 3)} + \dfrac{x + 3}{(x - 3)(x + 3)}$

$$= \dfrac{x^2 + x + 3}{(x - 3)(x + 3)}$$

44. $\dfrac{a}{2a - 2b} - \dfrac{b^2}{a^2 - b^2} = \dfrac{a}{2(a - b)} - \dfrac{b^2}{(a - b)(a + b)}$

$$= \dfrac{a(a + b)}{2(a - b)(a + b)} - \dfrac{2b^2}{2(a - b)(a + b)}$$

$$= \dfrac{a^2 + ab - 2b^2}{2(a - b)(a + b)}$$

$$= \dfrac{(a + 2b)(a - b)}{2(a - b)(a + b)}$$

$$= \dfrac{a + 2b}{2(a + b)}$$

46. $\dfrac{1}{x^2 + 3x - 10} + \dfrac{1}{x^2 - x - 2} = \dfrac{1}{(x + 5)(x - 2)} + \dfrac{1}{(x - 2)(x + 1)}$

$$= \dfrac{x + 1}{(x + 5)(x - 2)(x + 1)} + \dfrac{x + 5}{(x+5)(x-2)(x+1)}$$

$$= \dfrac{2x + 6}{(x + 5)(x - 2)(x + 1)}$$

48. $\dfrac{-4}{x^2 - 9} + \dfrac{1}{x - 3} - \dfrac{2}{3x + 9}$

$$= \dfrac{-4}{(x - 3)(x + 3)} + \dfrac{1}{x - 3} - \dfrac{2}{3(x + 3)}$$

$$= \dfrac{-12}{3(x - 3)(x + 3)} + \dfrac{3(x + 3)}{3(x - 3)(x + 3)} - \dfrac{2(x - 3)}{3(x - 3)(x + 3)}$$

$$= \dfrac{-12 + 3x + 9 - 2x + 6}{3(x - 3)(x + 3)}$$

$$= \dfrac{x + 3}{3(x - 3)(x + 3)}$$

$$= \dfrac{1}{3(x - 3)}$$

50. $1 + \dfrac{1}{a - 1} + \dfrac{a}{(a - 1)^2} = \dfrac{(a - 1)^2}{(a - 1)^2} + \dfrac{(a - 1)}{(a - 1)^2} + \dfrac{a}{(a - 1)^2}$

$$= \dfrac{a^2 - 2a + 1 + a - 1 + a}{(a - 1)^2}$$

$$= \dfrac{a^2}{(a - 1)^2}$$

Section 3.4

2. $\dfrac{1}{2} \cdot \dfrac{1}{2} = \dfrac{1 \cdot 1}{2 \cdot 2} = \dfrac{1}{4}$

4. $30 \cdot \dfrac{5}{6} = \dfrac{\overset{5}{\cancel{30}}}{1} \cdot \dfrac{5}{\cancel{6}_1} = \dfrac{25}{1} = 25$

6. $3 \cdot \dfrac{4}{5} \left(\dfrac{-6}{-7}\right)\left(\dfrac{-10}{21}\right) = \dfrac{\overset{1}{\cancel{3}}}{1} \cdot \dfrac{4}{\underset{1}{\cancel{5}}} \left(\dfrac{-6}{-7}\right)\left(\dfrac{\overset{-2}{\cancel{-10}}}{\underset{7}{\cancel{21}}}\right) = \dfrac{-48}{-49} = \dfrac{48}{49}$

8. $\dfrac{3x^3}{2y} \cdot \dfrac{7x}{4y} = \dfrac{21x^4}{8y^2}$

10. $\dfrac{-5x^5y}{8z^2} \cdot \dfrac{-4z}{15x^2y^4} = \dfrac{x^3}{6y^3z}$

12. $-\dfrac{4b}{15a^4} \cdot \dfrac{-3a^3}{14} \cdot \dfrac{-7a^2}{10b^2} = -\dfrac{a}{25b}$

$\dfrac{-12a^3b}{210a^4} , \dfrac{-7a^2}{10b^2} = \dfrac{84a^4b^2}{2100a^4b^2}$

14. $\dfrac{11a}{3a - 9} \cdot \dfrac{4a - 12}{22} = \dfrac{11a}{3(a - 3)} \cdot \dfrac{4(a - 3)}{22} = \dfrac{2a}{3}$

16. $\dfrac{a^2 - 4b^2}{a + 2} \cdot \dfrac{2 + a}{a + 2b} = \dfrac{(a - 2b)(a + 2b)}{a + 2} \cdot \dfrac{a + 2}{a + 2b} = a - 2b$

18. $\dfrac{x^2 + 3x - 4}{x^2 - 6x + 5} \cdot \dfrac{x - 5}{x^3 + 4x^2} = \dfrac{(x + 4)(x - 1)}{(x - 1)(x - 5)} \cdot \dfrac{x - 5}{x^2(x + 4)} = \dfrac{1}{x^2}$

20. $\dfrac{x^2 - y^2}{3x^2 + 11xy - 4y^2} \cdot \dfrac{x^2 + 3xy - 4y^2}{(x - y)^2}$

$= \dfrac{(x - y)(x + y)}{(3x - y)(x + 4y)} \cdot \dfrac{(x + 4y)(x - y)}{(x - y)(x - y)}$

$= \dfrac{x + y}{3x - y}$

22. $\dfrac{5}{6} \div \dfrac{7}{10} = \dfrac{5}{6} \cdot \dfrac{10}{7} = \dfrac{5}{\underset{3}{\cancel{6}}} \cdot \dfrac{\overset{5}{\cancel{10}}}{7} = \dfrac{25}{21}$

24. $\dfrac{1}{5} \div 2 = \dfrac{1}{5} \div \dfrac{2}{1} = \dfrac{1}{5} \cdot \dfrac{1}{2} = \dfrac{1}{10}$

26. $\dfrac{15b}{a} \div \dfrac{75a}{b} = \dfrac{15b}{a} \cdot \dfrac{b}{75a} = \dfrac{b^2}{5a^2}$

28. $\dfrac{14ac^3}{b^2c^3} \div \dfrac{-7}{a^4b^2c} = \dfrac{14ac^3}{b^2c^3} \cdot \dfrac{a^4b^2c}{-7} = -2a^5c$

30. $\dfrac{y}{x} \div \dfrac{xy}{xy - 1} = \dfrac{y}{x} \cdot \dfrac{xy - 1}{xy} = \dfrac{xy - 1}{x^2}$

32. $\dfrac{2x + 6}{x^2 + 9x} \div \dfrac{3x + 9}{x} = \dfrac{2(x + 3)}{x(x + 9)} \cdot \dfrac{x}{3(x + 3)} = \dfrac{2}{3(x + 9)}$

34. $\dfrac{1}{x^2 - 9} \div \dfrac{1}{x^2 + 9} = \dfrac{1}{x^2 - 9} \cdot \dfrac{x^2 + 9}{1} = \dfrac{x^2 + 9}{x^2 - 9}$

36. $\dfrac{x^2 + 4x - 12}{x^2 + 2x - 8} \div \dfrac{x^2 + 7x + 6}{x^2 - 1} = \dfrac{(x + 6)(x - 2)}{(x + 4)(x - 2)} \cdot \dfrac{(x - 1)(x + 1)}{(x + 1)(x + 6)}$

$$= \dfrac{x - 1}{x + 4}$$

38. $\dfrac{3a^2 + 7ab - 20b^2}{a^2 + 5ab + 4b^2} \div \dfrac{3a^2 - 17ab + 20b^2}{3a - 12b}$

$= \dfrac{(3a - 5b)(a + 4b)}{(a + b)(a + 4b)} \cdot \dfrac{3(a - 4b)}{(3a - 5b)(a - 4b)}$

$= \dfrac{3}{a + b}$

40. $\dfrac{a - 1}{a + 3} \div \dfrac{a^2 - 3a + 9}{a^3 + 27} = \dfrac{a - 1}{a + 3} \cdot \dfrac{(a + 3)(a^2 - 3a + 9)}{a^2 - 3a + 9} = a - 1$

42. $\left[\dfrac{1}{x + 5} - \dfrac{1}{x + 2}\right] \div \dfrac{3}{x^2 + 5x}$

$= \left[\dfrac{x + 2}{(x + 5)(x + 2)} - \dfrac{x + 5}{(x + 5)(x + 2)}\right] \cdot \dfrac{x(x + 5)}{3}$

$= \left[\dfrac{-3}{(x + 5)(x + 2)}\right] \cdot \dfrac{x(x + 5)}{3}$

$= -\dfrac{x}{x + 2}$

Section 3.5

2. $\dfrac{32x^3 y^3}{8x^2 y^3} = 4x$

4. $\dfrac{7x + 7y}{7} = \dfrac{7x}{7} + \dfrac{7y}{7} = x + y$

6. $\dfrac{8x^2 - 12x + 4}{4} = \dfrac{8x^2}{4} - \dfrac{12x}{4} + \dfrac{4}{4} = 2x^2 - 3x + 1$

8. $\dfrac{10a^2 + 15a - 1}{5} = \dfrac{10a^2}{5} + \dfrac{15a}{5} - \dfrac{1}{5} = 2a^2 + 3a - \dfrac{1}{5}$

10. $\dfrac{24t^3 - 6t^2 + 18t}{-6t} = \dfrac{24t^3}{-6t} - \dfrac{6t^2}{-6t} + \dfrac{18t}{-6t} = -4t^2 + t - 3$

12. $\dfrac{r^4 s^2 - r^3 s^3 - r^2 s^2}{r^2 s^2} = \dfrac{r^4 s^2}{r^2 s^2} - \dfrac{r^3 s^3}{r^2 s^2} - \dfrac{r^2 s^2}{r^2 s^2} = r^2 - rs - 1$

24

14. $\dfrac{80x^3y^4 + 12x^3y^3 + 48x^2y^4}{16x^2y^2} = \dfrac{80x^3y^4}{16x^2y^2} + \dfrac{12x^3y^3}{16x^2y^2} + \dfrac{48x^2y^4}{16x^2y^2}$

$\qquad\qquad\qquad\qquad = 5xy^2 + \dfrac{3}{4}xy + 3y^2$

16. $\dfrac{50x^2y^2 - 75xy + 5}{25xy} = \dfrac{50x^2y^2}{25xy} - \dfrac{75xy}{25xy} + \dfrac{5}{25xy} = 2xy - 3 + \dfrac{1}{5xy}$

18. $\dfrac{h^3 - 4h^2 + 7h - 1}{h} = \dfrac{h^3}{h} - \dfrac{4h^2}{h} + \dfrac{7h}{h} - \dfrac{1}{h} = h^2 - 4h + 7 - \dfrac{1}{h}$

20.
```
            x  + 7
      _____
x + 2 | x² + 9x + 14
        x² + 2x
        _____
             7x + 14
             7x + 14
             _____
                   0
```

22.
```
            2x + 5
      _____
x - 4 | 2x² - 3x - 25
        2x² - 8x
        _____
              5x - 25
              5x - 20
              _____
                 - 5
```

24.
```
             x² + 3x  - 1
       _____
2x - 1 | 2x³ + 5x² - 5x + 1
         2x³ -  x²
         _____
               6x² - 5x
               6x² - 3x
               _____
                   - 2x + 1
                   - 2x + 1
                   _____
                         0
```

26.
```
             x² -  x  + 5
       _____
4x + 3 | 4x³ -  x² + 17x - 15
         4x³ + 3x²
         _____
             - 4x² + 17x
             - 4x² -  3x
             _____
                     20x - 15
                     20x + 15
                     _____
                        - 30
```

28.
```
            x² + 3x  + 4
      _____
x - 3 | x³       - 5x - 12
        x³ - 3x²
        _____
             3x² - 5x
             3x² - 9x
             _____
                  4x - 12
                  4x - 12
                  _____
                        0
```

30.
```
               3x  + 4
          _____
x² + x - 2 | 3x³ + 7x² -  x - 1
             3x³ + 3x² - 6x
             _____
                   4x² + 5x - 1
                   4x² + 4x - 8
                   _____
                        x + 7
```

25

32.

$$
\begin{array}{r}
3x^2 + x - 3 \\
2x^2 - 4x + 3 \enclose{longdiv}{6x^4 - 10x^3 - x^2 \qquad - 11} \\
\underline{6x^4 - 12x^3 + 9x^2} \\
2x^3 - 10x^2 \\
\underline{2x^3 - 4x^2 + 3x} \\
-6x^2 - 3x - 11 \\
\underline{-6x^2 + 12x - 9} \\
-15x - 2
\end{array}
$$

34.

$$
\begin{array}{r}
x \\
x^2 - 1 \enclose{longdiv}{x^3 \qquad - 1} \\
\underline{x^3 - x} \\
x - 1
\end{array}
$$

36.

$$
\begin{array}{r}
x^2 + 2x - 8 \\
x + 3 \enclose{longdiv}{x^3 + 5x^2 - 2x + k} \\
\underline{x^3 + 3x^2} \\
2x^2 - 2x \\
\underline{2x^2 + 6x} \\
-8x + k \\
\underline{-8x - 24}
\end{array}
$$

k = -24 will give a remainder of 0.

Section 3.6

2. $\dfrac{\frac{3}{4}}{5} = \dfrac{\frac{3}{4} \cdot 4}{5 \cdot 4} = \dfrac{3}{20}$

4. $\dfrac{-\frac{a}{b}}{\frac{a}{c}} = \dfrac{(-\frac{a}{b})bc}{(\frac{a}{c})bc} = \dfrac{-ac}{a^3 b} = -\dfrac{c}{a^2 b}$

6. $\dfrac{1 + \frac{1}{2}}{5} = \dfrac{(1 + \frac{1}{2}) \cdot 2}{5 \cdot 2} = \dfrac{2 + 1}{10} = \dfrac{3}{10}$

8. $\dfrac{\frac{1}{4} + \frac{1}{6}}{\frac{1}{2} + \frac{1}{3}} = \dfrac{(\frac{1}{4} + \frac{1}{6})12}{(\frac{1}{2} + \frac{1}{3})12} = \dfrac{3 + 2}{6 + 4} = \dfrac{5}{10} = \dfrac{1}{2}$

10. $\dfrac{\frac{a}{b} + 1}{\frac{a}{b} - 1} = \dfrac{(\frac{a}{b} + 1)b}{(\frac{a}{b} - 1)b} = \dfrac{a + b}{a - b}$

12. $\dfrac{\frac{3}{x} + 1}{1 + \frac{3}{y}} = \dfrac{(\frac{3}{x} + 1)xy}{(1 + \frac{3}{y})xy} = \dfrac{3y + xy}{xy + 3x}$

14. $$\frac{9 - \dfrac{1}{x^2}}{3 + \dfrac{1}{x}} = \frac{(9 - \dfrac{1}{x^2})x^2}{(3 + \dfrac{1}{x})x^2} = \frac{9x^2 - 1}{3x^2 + x} = \frac{(3x - 1)(3x + 1)}{x(3x + 1)} = \frac{3x - 1}{x}$$

16. $$\frac{\dfrac{1}{b} + \dfrac{1}{ab}}{\dfrac{1}{b} + \dfrac{1}{a}} = \frac{(\dfrac{1}{b} + \dfrac{1}{ab})ab}{(\dfrac{1}{b} + \dfrac{1}{a})ab} = \frac{a + 1}{a + b}$$

18. $$\frac{\dfrac{x}{9} - \dfrac{1}{x}}{1 + \dfrac{x + 6}{x}} = \frac{(\dfrac{x}{9} - \dfrac{1}{x})9x}{(1 + \dfrac{x + 6}{x})9x}$$

$$= \frac{x^2 - 9}{9x + 9(x + 6)}$$

$$= \frac{(x - 3)(x + 3)}{18x + 54}$$

$$= \frac{(x - 3)(x + 3)}{18(x + 3)}$$

$$= \frac{x - 3}{18}$$

20. $$\frac{1 - \dfrac{7}{x + 1}}{\dfrac{4}{1 + x} + 1} = \frac{(1 - \dfrac{7}{x + 1})(x + 1)}{(\dfrac{4}{x + 1} + 1)(x + 1)} = \frac{(x + 1) - 7}{4 + (x + 1)} = \frac{x - 6}{x + 5}$$

22. $$\frac{\dfrac{1}{y^3} + \dfrac{1}{x^3}}{x^2 - xy + y^2} = \frac{(\dfrac{1}{y^3} + \dfrac{1}{x^3})x^3 y^3}{(x^2 - xy + y^2)x^3 y^3}$$

$$= \frac{x^3 + y^3}{(x^2 - xy + y^2)x^3 y^3}$$

$$= \frac{(x + y)(x^2 - xy + y^2)}{(x^2 - xy + y^2)x^3 y^3}$$

$$= \frac{x + y}{x^3 y^3}$$

24. $$\dfrac{\dfrac{1}{(x+h)^2} - \dfrac{1}{x^2}}{h} = \dfrac{[\dfrac{1}{(x+h)^2} - \dfrac{1}{x^2}](x+h)^2 x^2}{h(x+h)^2 x^2} = \dfrac{x^2 - (x+h)^2}{h(x+h)^2 x^2}$$

$$= \dfrac{x^2 - x^2 - 2hx - h^2}{h(x+h)^2 x^2}$$

$$= \dfrac{h(-2x - h)}{h(x+h)^2 x^2}$$

$$= \dfrac{-2x - h}{(x+h)^2 x^2}$$

26. $$1 - \dfrac{1}{1 - \dfrac{1}{1 - \dfrac{1}{x}}} = 1 - \dfrac{1}{1 - \dfrac{1 \cdot x}{(1 - \dfrac{1}{x}) \cdot x}}$$

$$= 1 - \dfrac{1}{1 - \dfrac{x}{x - 1}}$$

$$= 1 - \dfrac{1 \cdot (x - 1)}{(1 - \dfrac{x}{x - 1}) \cdot (x - 1)}$$

$$= 1 - \dfrac{x - 1}{(x - 1) - x}$$

$$= 1 - \dfrac{x - 1}{-1}$$

$$= 1 + x - 1$$

$$= x$$

CHAPTER FOUR
EXPONENTS, ROOTS, AND RADICALS

Section 4.1

2. $999^0 = 1$

4. $(8xy^2)^0 = 1$

6. $(-3)^0 = 1$ and $-3^0 = -1$

8. $2^{-1} = \dfrac{1}{2^1} = \dfrac{1}{2}$

10. $5^{-3} = \dfrac{1}{5^3} = \dfrac{1}{125}$

12. $(\dfrac{1}{3})^{-4} = (\dfrac{3}{1})^4 = 81$

14. $20(\dfrac{4}{5})^{-2} = 20(\dfrac{5}{4})^2 = 20(\dfrac{25}{16}) = \dfrac{125}{4}$

16. $\dfrac{1}{10^{-3}} = 10^3 = 1000$

18. $8 \cdot 4^{-2} = 8 \cdot \dfrac{1}{4^2} = 8 \cdot \dfrac{1}{16} = \dfrac{1}{2}$

20. $\dfrac{5y^0}{6^{-2}} = \dfrac{5}{6^{-2}} = 5 \cdot 6^2 = 5 \cdot 36 = 180$

22. $4^2 + 4^{-2} = 4^2 + \dfrac{1}{4^2} = 16 + \dfrac{1}{16} = \dfrac{256}{16} + \dfrac{1}{16} = \dfrac{257}{16}$

24. $2^{-1} + 5^{-1} = \dfrac{1}{2} + \dfrac{1}{5} = \dfrac{5}{10} + \dfrac{2}{10} = \dfrac{7}{10}$

26. $\dfrac{3^{-2}}{4^{-2}} = \dfrac{4^2}{3^2} = \dfrac{16}{9}$

28. $x^{-4}x^6 = x^{-4+6} = x^2$

30. $x^2 x^{-8} = x^{2+(-8)} = x^{-6} = \dfrac{1}{x^6}$

32. $\dfrac{x^3}{x^{-3}} = x^{3-(-3)} = x^{3+3} = x^6$

34. $\dfrac{x^{-1}}{x^5 y^0} = \dfrac{x^{-1}}{x^5} = x^{-1-5} = x^{-6} = \dfrac{1}{x^6}$

36. $(x^{-4})^5 = x^{-4 \cdot 5} = x^{-20} = \dfrac{1}{x^{20}}$

38. $(x^{-2}y^3)^{-2} = (x^{-2})^{-2}(y^3)^{-2} = x^4 y^{-6} = \dfrac{x^4}{y^6}$

40. $(5xy^{-3})^{-1} = 5^{-1}x^{-1}(y^{-3})^{-1} = 5^{-1}x^{-1}y^3 = \dfrac{y^3}{5x}$

42. $\dfrac{2a^{-3}}{5b^{-3}} = \dfrac{2b^3}{5a^3}$

44. $\left[\left(\dfrac{a^2b^{-1}}{c^{-3}}\right)^{-2}\right]^{-1} = \left(\dfrac{a^2b^{-1}}{c^{-3}}\right)^2 = \dfrac{a^4b^{-2}}{c^{-6}} = \dfrac{a^4c^6}{b^2}$

46. $(x - y)^{-1} = \dfrac{1}{(x-y)^1} = \dfrac{1}{x - y}$

48. $x^{-2} + y^{-2} = \dfrac{1}{x^2} + \dfrac{1}{y^2} = \dfrac{y^2}{x^2y^2} + \dfrac{x^2}{x^2y^2} = \dfrac{y^2 + x^2}{x^2y^2}$

50. $(x^{-2} + y^{-2})^{-1} = \dfrac{1}{x^{-2} + y^{-2}} = \dfrac{1}{\dfrac{1}{x^2} + \dfrac{1}{y^2}} = \dfrac{1 \cdot x^2y^2}{\left(\dfrac{1}{x^2} + \dfrac{1}{y^2}\right) \cdot x^2y^2}$

$$= \dfrac{x^2y^2}{y^2 + x^2}$$

52. $\dfrac{x^{-1} - y^{-1}}{x^{-1} + y^{-1}} = \dfrac{\dfrac{1}{x} - \dfrac{1}{y}}{\dfrac{1}{x} + \dfrac{1}{y}} = \dfrac{\left(\dfrac{1}{x} - \dfrac{1}{y}\right)xy}{\left(\dfrac{1}{x} + \dfrac{1}{y}\right)xy} = \dfrac{y - x}{y + x}$

54. $\dfrac{x^{-1} - y^{-1}}{(xy)^{-1}} = \dfrac{\dfrac{1}{x} - \dfrac{1}{y}}{\dfrac{1}{xy}} = \dfrac{\left(\dfrac{1}{x} - \dfrac{1}{y}\right)xy}{\left(\dfrac{1}{xy}\right)xy} = \dfrac{y - x}{1} = y - x$

56. $(x - y)(x^{-1} + y^{-1}) = xx^{-1} + xy^{-1} - yx^{-1} - yy^{-1} = 1 + \dfrac{x}{y} - \dfrac{y}{x} - 1$

$$= \dfrac{x^2}{xy} - \dfrac{y^2}{xy}$$

$$= \dfrac{x^2 - y^2}{xy}$$

58. $\left(\dfrac{a}{b}\right)^n = \underbrace{\dfrac{a}{b} \cdot \dfrac{a}{b} \cdots \dfrac{a}{b}}_{n \text{ factors of } \frac{a}{b}} = \dfrac{\overbrace{a \cdot a \cdots a}^{n \text{ factors}}}{\underbrace{b \cdot b \cdots b}_{n \text{ factors}}} = \dfrac{a^n}{b^n}$

Section 4.2

2. 200   4. 3,100,000   6. 0.0045   8. 0.000909

10. 0.000008819   12. 700,000,000

14. 7×10^3 16. 1.4×10^6 18. 8.1×10^{-3} 20. 5.95×10^{-6}

22. 4.31×10^{11} 24. 6.075×10^{-1}

26. $(0.0003)(20,000)$

$= (3 \times 10^{-4})(2 \times 10^4)$

$= (3 \cdot 2) \times (10^{-4} 10^4)$

$= 6 \times 10^0$

$= 6$

28. $(6,000,000)(0.000004)$

$= (6 \times 10^6)(4 \times 10^{-6})$

$= (6 \cdot 4) \times (10^6 10^{-6})$

$= 24 \times 10^0$

$= 24$

$= 2.4 \times 10$

30. $\dfrac{0.000081}{0.0009}$

$= \dfrac{8.1 \times 10^{-5}}{9 \times 10^{-4}}$

$= \dfrac{8.1}{9} \times \dfrac{10^{-5}}{10^{-4}}$

$= 0.9 \times 10^{-1}$

$= 9 \times 10^{-1} \times 10^{-1}$

$= 9 \times 10^{-2}$

32. $\dfrac{0.000144}{0.00000012}$

$= \dfrac{1.44 \times 10^{-4}}{1.2 \times 10^{-7}}$

$= \dfrac{1.44}{1.2} \times \dfrac{10^{-4}}{10^{-7}}$

$= 1.2 \times 10^3$

34. $\dfrac{(6600)(0.00225)}{(0.00015)(11,000,000)}$

$= \dfrac{(6.6 \times 10^3)(2.25 \times 10^{-3})}{(1.5 \times 10^{-4})(1.1 \times 10^{+7})}$

$= \dfrac{(6.6)(2.25)}{(1.5)(1.1)} \times \dfrac{10^3 10^{-3}}{10^{-4} 10^{+7}}$

$= 9 \times 10^{11}$

36. $\dfrac{(0.0036)(3,000,000)}{(40,000)(0.0006)}$

$= \dfrac{(3.6 \times 10^{-3})(3 \times 10^6)}{(4 \times 10^4)(6 \times 10^{-4})}$

$= \dfrac{(3.6)(3)}{(4)(6)} \times \dfrac{10^{-3} 10^6}{10^4 10^{-4}}$

$= 0.45 \times 10^3$

$= 4.5 \times 10^{-1} \times 10^3$

$= 4.5 \times 10^2$

Section 4.3

2. 6   4. -11   6. ± 10 8. 0.2

10. $\dfrac{1}{8}$ 12. $\dfrac{2}{13}$ 14. 6   16. -3

18. $-(-3) = 3$ 20. 3   22. 2   24. -1

26. ab^3 28. $-12a^5$ 30. x^4 32. $-3x^2y^2$

34. $-x^2y^2$ 36. $-xy^5$ 38. $(x + y)^2$ 40. x^ny^m

42. $\sqrt{x^2 + 2x + 1} = \sqrt{(x + 1)^2} = x + 1$

Section 4.4

2. $27^{2/3} = (\sqrt[3]{27})^2 = (3)^2 = 9$

4. $25^{3/2} = (\sqrt{25})^3 = (5)^3 = 125$

6. $49^{1/2} = (\sqrt{49})^1 = (7)^1 = 7$

8. $(-32)^{3/5} = (\sqrt[5]{-32})^3 = (-2)^3 = -8$

10. $(-1)^{4/3} = (\sqrt[3]{-1})^4 = (-1)^4 = 1$

12. $243^{1/5} = (\sqrt[5]{243})^1 = 3$

14. $4^{-(1/2)} = \dfrac{1}{4^{1/2}} = \dfrac{1}{\sqrt{4}} = \dfrac{1}{2}$

16. $27^{-(2/3)} = \dfrac{1}{27^{2/3}} = \dfrac{1}{(\sqrt[3]{27})^2} = \dfrac{1}{(3)^2} = \dfrac{1}{9}$

18. $81^{-(1/4)} = \dfrac{1}{81^{1/4}} = \dfrac{1}{\sqrt[4]{81}} = \dfrac{1}{3}$

20. $(-64)^{-(4/3)} = \dfrac{1}{(-64)^{4/3}} = \dfrac{1}{(\sqrt[3]{-64})^4} = \dfrac{1}{(-4)^4} = \dfrac{1}{256}$

22. $x^{1/4} \cdot x^{1/4} = x^{(1/4)+(1/4)} = x^{1/2}$

24. $x^{-(1/3)} \cdot x^{5/3} = x^{-(1/3)+(5/3)} = x^{4/3}$

26. $\dfrac{x^{5/2}}{x^{1/2}} = x^{(5/2)-(1/2)} = x^{4/2} = x^2$

28. $\dfrac{x^{-(1/2)}y^{1/2}y^{1/5}}{xy^{1/5}} = \dfrac{x^{-(1/2)}y^{1/2}}{x} = \dfrac{y^{1/2}}{x^{1/2}x} = \dfrac{y^{1/2}}{x^{3/2}}$

30. $\dfrac{x^{-(1/2)}}{y^{-1}} \div \dfrac{y^{4/5}}{x^{1/2}} = \dfrac{x^{-(1/2)}}{y^{-1}} \cdot \dfrac{x^{1/2}}{y^{4/5}} = \dfrac{x^0}{y^{-(1/5)}} = y^{1/5}$

32. $x^{-(1/3)}(x^{1/3} + x) = x^{-(1/3)}x^{1/3} + x^{-(1/3)}x = 1 + x^{2/3}$

34. $(x^{1/2} + y^{1/2})(x^{1/2} - y^{1/2}) = x^{1/2}x^{1/2} - x^{1/2}y^{1/2} + y^{1/2}x^{1/2} - y^{1/2}y^{1/2}$

$$= x - y$$

36. $\dfrac{1}{x^{1/3}} - \dfrac{1}{y^{1/3}} = \dfrac{y^{1/3}}{x^{1/3}y^{1/3}} - \dfrac{x^{1/3}}{x^{1/3}y^{1/3}} = \dfrac{y^{1/3} - x^{1/3}}{x^{1/3}y^{1/3}}$

38. $(4)^{1/4} = (2^2)^{1/4} = 2^{2 \cdot (1/4)} = 2^{1/2}$

40. $(x^6)^{1/3} = x^{6 \cdot (1/3)} = x^2$

42. $(a^{1/3}b^{1/2})^6 = (a^{1/3})^6 (b^{1/2})^6 = a^2 b^3$

44. $(a^{5/2}b^{1/3})^{-6}(a^2 b^{-(1/2)})^{-1} = (a^{5/2})^{-6}(b^{1/3})^{-6}(a^2)^{-1}(b^{-(1/2)})^{-1}$

$$= a^{-15}b^{-2}a^{-2}b^{1/2}$$

$$= a^{-17}b^{-(3/2)}$$

$$= \dfrac{1}{a^{17}b^{3/2}}$$

46. $\left(\dfrac{a^{2/3}b^{-3/4}}{c^{5/12}}\right)^{12} = \dfrac{(a^{2/3})^{12}(b^{-(3/4)})^{12}}{(c^{5/12})^{12}} = \dfrac{a^8 b^{-9}}{c^5} = \dfrac{a^8}{b^9 c^5}$

48. $\left[\left(\dfrac{x^6(x+3)^3}{27}\right)^{1/3}\right]^{-1} = \left(\dfrac{x^6(x+3)^3}{27}\right)^{-(1/3)}$

$$= \left(\dfrac{27}{x^6(x+3)^3}\right)^{1/3}$$

$$= \dfrac{3}{x^2(x+3)}$$

50. $\left[\left(\dfrac{x^{n/2}}{y^{n/3}}\right)^6\right]^{1/n} = \left[\dfrac{x^{3n}}{y^{2n}}\right]^{1/n} = \dfrac{x^3}{y^2}$

52. $\sqrt[6]{x^3} = (x^3)^{1/6} = x^{3 \cdot (1/6)} = x^{1/2}$

54. $\sqrt[12]{\sqrt[3]{x}} = (x^{1/3})^{1/12} = x^{(1/3) \cdot (1/12)} = x^{1/36}$

56. $\sqrt[3]{\sqrt{x}} = (x^{1/2})^{1/3} = x^{(1/2) \cdot (1/3)} = x^{1/6}$

58. $\sqrt{\sqrt[6]{x}} = (x^{1/6})^{1/2} = x^{(1/6) \cdot (1/2)} = x^{1/12}$

60. $\dfrac{1}{\sqrt[3]{x}} = \dfrac{1}{x^{1/3}} = x^{-(1/3)}$

62. $\dfrac{10}{\sqrt{(x+1)^3}} = \dfrac{10}{(x+1)^{3/2}} = 10(x+1)^{-(3/2)}$

Section 4.5

2. $\sqrt{12} = \sqrt{4 \cdot 3} = \sqrt{4} \cdot \sqrt{3} = 2 \cdot \sqrt{3} = 2\sqrt{3}$

4. $\sqrt{300} = \sqrt{100 \cdot 3} = \sqrt{100}\sqrt{3} = 10\sqrt{3}$

6. $-\sqrt{48} = -\sqrt{16 \cdot 3} = -\sqrt{16}\sqrt{3} = -4\sqrt{3}$

8. $\sqrt[3]{54} = \sqrt[3]{27 \cdot 2} = \sqrt[3]{27}\sqrt[3]{2} = 3\sqrt[3]{2}$

10. $\sqrt[5]{160} = \sqrt[5]{32 \cdot 5} = \sqrt[5]{32}\sqrt[5]{5} = 2\sqrt[5]{5}$

12. $\sqrt{18x^3y^2} = \sqrt{9 \cdot 2x^2 xy^2} = \sqrt{9x^2y^2}\sqrt{2x} = 3xy\sqrt{2x}$

14. $\sqrt[3]{64x^8y^8} = \sqrt[3]{64x^6x^2y^6y^2} = \sqrt[3]{64x^6y^6}\sqrt[3]{x^2y^2} = 4x^2y^2\sqrt[3]{x^2y^2}$

16. $\pm\sqrt[4]{3^4x^4y^5} = \pm\sqrt[4]{3^4x^4y^4y} = \pm\sqrt[4]{3^4x^4y^4}\sqrt[4]{y} = \pm 3xy\sqrt[4]{y}$

18. $\sqrt{3}\sqrt{27} = \sqrt{3 \cdot 27} = \sqrt{81} = 9$

20. $(\sqrt{5})^2 = \sqrt{5}\sqrt{5} = \sqrt{5 \cdot 5} = \sqrt{25} = 5$

22. $(\sqrt{3})^3 = \sqrt{3}\sqrt{3}\sqrt{3} = 3\sqrt{3}$

24. $\sqrt{3x}\sqrt{15xy} = \sqrt{3x \cdot 15xy} = \sqrt{45x^2y} = \sqrt{9 \cdot 5x^2y} = 3x\sqrt{5y}$

26. $\sqrt{x^2y}\sqrt{xy^5} = \sqrt{x^3y^6} = \sqrt{x^2xy^6} = xy^3\sqrt{x}$

28. $\sqrt[3]{2xy^5}\sqrt[3]{4xy} = \sqrt[3]{8x^2y^6} = 2y^2\sqrt[3]{x^2}$

30. $\sqrt{\dfrac{4}{49}} = \dfrac{\sqrt{4}}{\sqrt{49}} = \dfrac{2}{7}$

32. $\sqrt{\dfrac{5}{36}} = \dfrac{\sqrt{5}}{\sqrt{36}} = \dfrac{\sqrt{5}}{6}$

34. $\sqrt[3]{\dfrac{54}{125}} = \dfrac{\sqrt[3]{54}}{\sqrt[3]{125}} = \dfrac{\sqrt[3]{27 \cdot 2}}{5} = \dfrac{3\sqrt[3]{2}}{5}$

36. $\sqrt{\dfrac{63xy^3}{y^5}} = \sqrt{\dfrac{63x}{y^2}} = \sqrt{\dfrac{9 \cdot 7x}{y^2}} = \dfrac{3\sqrt{7x}}{y}$

38. $\dfrac{\sqrt{90}}{\sqrt{10}} = \sqrt{\dfrac{90}{10}} = \sqrt{9} = 3$

40. $\dfrac{\sqrt[3]{x^5y^6}}{\sqrt[3]{x^2y^2}} = \sqrt[3]{\dfrac{x^5y^6}{x^2y^2}} = \sqrt[3]{x^3y^4} = \sqrt[3]{x^3y^3y} = xy\sqrt[3]{y}$

42. $\dfrac{\sqrt{48x^3y^3}}{\sqrt{3xy}} = \sqrt{\dfrac{48x^3y^3}{3xy}} = \sqrt{16x^2y^2} = 4xy$

44. $\dfrac{\sqrt{x^{11}y^8z^5}}{\sqrt{x^2yz^3}} = \sqrt{\dfrac{x^{11}y^8z^5}{x^2yz^3}} = \sqrt{x^9y^7z^2} = x^4y^3z\sqrt{xy}$

46. $\dfrac{1}{\sqrt{5}} = \dfrac{1 \cdot \sqrt{5}}{\sqrt{5} \cdot \sqrt{5}} = \dfrac{\sqrt{5}}{5}$ 48. $\dfrac{3}{\sqrt{7}} = \dfrac{3 \cdot \sqrt{7}}{\sqrt{7} \cdot \sqrt{7}} = \dfrac{3\sqrt{7}}{7}$

50. $\dfrac{6}{\sqrt{2}} = \dfrac{6 \cdot \sqrt{2}}{\sqrt{2} \cdot \sqrt{2}} = \dfrac{6\sqrt{2}}{2} = 3\sqrt{2}$

52. $\sqrt{\dfrac{7}{10}} = \dfrac{\sqrt{7}}{\sqrt{10}} = \dfrac{\sqrt{7} \cdot \sqrt{10}}{\sqrt{10} \cdot \sqrt{10}} = \dfrac{\sqrt{70}}{10}$

54. $\dfrac{6}{\sqrt{3a}} = \dfrac{6 \cdot \sqrt{3a}}{\sqrt{3a} \cdot \sqrt{3a}} = \dfrac{6\sqrt{3a}}{3a} = \dfrac{2\sqrt{3a}}{a}$

56. $\dfrac{4}{\sqrt[3]{3}} = \dfrac{4 \cdot \sqrt[3]{9}}{\sqrt[3]{3} \cdot \sqrt[3]{9}} = \dfrac{4\sqrt[3]{9}}{\sqrt[3]{27}} = \dfrac{4\sqrt[3]{9}}{3}$

58. $\dfrac{1}{\sqrt{x^3}} = \dfrac{1 \cdot \sqrt{x}}{\sqrt{x^3} \cdot \sqrt{x}} = \dfrac{\sqrt{x}}{\sqrt{x^4}} = \dfrac{\sqrt{x}}{x^2}$

60. $\dfrac{1}{\sqrt[3]{(x+1)^2}} = \dfrac{1 \cdot \sqrt[3]{x+1}}{\sqrt[3]{(x+1)^2} \cdot \sqrt[3]{x+1}} = \dfrac{\sqrt[3]{x+1}}{\sqrt[3]{(x+1)^3}} = \dfrac{\sqrt[3]{x+1}}{x+1}$

62. $\sqrt{x^2y^2 + x^3y^3} = \sqrt{x^2y^2(1+xy)} = xy\sqrt{1+xy}$

64. $\sqrt{a+b}\sqrt{a^2+2ab+b^2} = \sqrt{a+b}\ \sqrt{(a+b)^2} = \sqrt{a+b}\ (a+b)$

66. $\dfrac{\sqrt{a^{3n}(a+b)^3}}{\sqrt{a^n(a+b)}} = \sqrt{a^{2n}(a+b)^2} = a^n(a+b)$

68. $\sqrt{25-16} = \sqrt{9} = 3$ but $\sqrt{25} - \sqrt{16} = 5 - 4 = 1$.

Section 4.6

2. $4\sqrt{5} + 3\sqrt{5} = (4+3)\sqrt{5} = 7\sqrt{5}$

4. $\sqrt{3} - 8\sqrt{3} + 7\sqrt{7} = (1-8)\sqrt{3} + 7\sqrt{7} = -7\sqrt{3} + 7\sqrt{7}$

6. $10\sqrt{xy+1} + 3\sqrt{xy+1} = (10+3)\sqrt{xy+1} = 13\sqrt{xy+1}$

8. $\sqrt{8} + \sqrt{18} = 2\sqrt{2} + 3\sqrt{2} = 5\sqrt{2}$

10. $\sqrt{75x} - \sqrt{12x} = 5\sqrt{3x} - 2\sqrt{3x} = 3\sqrt{3x}$

12. $4\sqrt{11xy^2} + 5\sqrt{99xy^2} = 4y\sqrt{11x} + 5 \cdot 3y\sqrt{11x} = 19y\sqrt{11x}$

14. $6\sqrt{8a} + 2\sqrt{50a} - 3\sqrt{2a} = 6 \cdot 2\sqrt{2a} + 2 \cdot 5\sqrt{2a} - 3\sqrt{2a} = 19\sqrt{2a}$

16. $\sqrt[3]{24a} - 2\sqrt[3]{3a} + \sqrt[3]{192a} = 2\sqrt[3]{3a} - 2\sqrt[3]{3a} + 4\sqrt[3]{3a} = 4\sqrt[3]{3a}$

18. $\sqrt{48} + \sqrt{\dfrac{1}{3}}$

$= 4\sqrt{3} + \dfrac{1}{\sqrt{3}}$

$= \dfrac{4\sqrt{3}\sqrt{3}}{\sqrt{3}} + \dfrac{1}{\sqrt{3}}$

$= \dfrac{12}{\sqrt{3}} + \dfrac{1}{\sqrt{3}}$

$= \dfrac{13}{\sqrt{3}}$

$= \dfrac{13\sqrt{3}}{3}$

20. $\sqrt{90} + \sqrt{\dfrac{2}{5}}$

$= 3\sqrt{10} + \dfrac{\sqrt{2}}{\sqrt{5}}$

$= \dfrac{3\sqrt{10}\sqrt{5}}{\sqrt{5}} + \dfrac{\sqrt{2}}{\sqrt{5}}$

$= \dfrac{3\sqrt{50}}{\sqrt{5}} + \dfrac{\sqrt{2}}{\sqrt{5}}$

$= \dfrac{3 \cdot 5\sqrt{2}}{\sqrt{5}} + \dfrac{\sqrt{2}}{\sqrt{5}}$

$= \dfrac{16\sqrt{2}}{\sqrt{5}}$

$= \dfrac{16\sqrt{10}}{5}$

22. $7(\sqrt{2} + 1) = 7 \cdot \sqrt{2} + 7 \cdot 1 = 7\sqrt{2} + 7$

24. $3(4\sqrt{13} - 5) = 3 \cdot 4\sqrt{13} - 3 \cdot 5 = 12\sqrt{13} - 15$

26. $\sqrt{5}(\sqrt{2} + \sqrt{5}) = \sqrt{5}\ \sqrt{2} + \sqrt{5}\sqrt{5} = \sqrt{10} + 5$

28. $\sqrt{x}(\sqrt{xy} + \sqrt{x}) = \sqrt{x}\ \sqrt{xy} + \sqrt{x}\sqrt{x} = \sqrt{x^2 y} + x = x\sqrt{y} + x$

30. $(4 + \sqrt{3})(1 - \sqrt{3}) = 4 - 4\sqrt{3} + \sqrt{3} - \sqrt{3}\sqrt{3} = 1 - 3\sqrt{3}$

32. $(\sqrt{x} - 1)(\sqrt{x} + 1) = \sqrt{x}\sqrt{x} + \sqrt{x} - \sqrt{x} - 1 = x - 1$

34. $(\sqrt{3} + \sqrt{7})(\sqrt{3} - \sqrt{7}) = \sqrt{3}\sqrt{3} - \sqrt{3}\sqrt{7} + \sqrt{7}\sqrt{3} - \sqrt{7}\ \sqrt{7} = -4$

36. $(\sqrt{10x} - 2\sqrt{5x})^2 = 10x - 4\sqrt{50x^2} + 4(5x)$

$= 10x - 4 \cdot 5x\sqrt{2} + 20x$

$= 30x - 20x\sqrt{2}$

38. $\dfrac{6}{3 + \sqrt{3}}$

$= \dfrac{6(3 - \sqrt{3})}{(3 + \sqrt{3})(3 - \sqrt{3})}$

$= \dfrac{6(3 - \sqrt{3})}{9 - 3\sqrt{3} + 3\sqrt{3} - 3}$

$= \dfrac{6(3 - \sqrt{3})}{6}$

$= 3 - \sqrt{3}$

40. $\dfrac{\sqrt{2}}{\sqrt{5} - \sqrt{2}}$

$= \dfrac{\sqrt{2}(\sqrt{5} + \sqrt{2})}{(\sqrt{5} - \sqrt{2})(\sqrt{5} + \sqrt{2})}$

$= \dfrac{\sqrt{10} + 2}{5 + \sqrt{10} - \sqrt{10} - 2}$

$= \dfrac{\sqrt{10} + 2}{3}$

42. $\dfrac{\sqrt{5} - \sqrt{3}}{\sqrt{5} + \sqrt{3}}$

$= \dfrac{(\sqrt{5} - \sqrt{3})(\sqrt{5} - \sqrt{3})}{(\sqrt{5} + \sqrt{3})(\sqrt{5} - \sqrt{3})}$

$= \dfrac{5 - \sqrt{15} - \sqrt{15} + 3}{5 - \sqrt{15} + \sqrt{15} - 3}$

$= \dfrac{8 - 2\sqrt{15}}{2}$

$= 4 - \sqrt{15}$

44. $\dfrac{1}{\sqrt{x}} - \dfrac{1}{\sqrt{y}}$

$= \dfrac{\sqrt{x}}{x} - \dfrac{\sqrt{y}}{y}$

$= \dfrac{y\sqrt{x}}{xy} - \dfrac{x\sqrt{y}}{xy}$

$= \dfrac{y\sqrt{x} - x\sqrt{y}}{xy}$

46. $\dfrac{1}{\sqrt{x + 5} + \sqrt{x}} = \dfrac{1(\sqrt{x + 5} - \sqrt{x})}{(\sqrt{x + 5} + \sqrt{x})(\sqrt{x + 5} - \sqrt{x})}$

$= \dfrac{\sqrt{x + 5} - \sqrt{x}}{(x + 5) - \sqrt{x + 5}\,\sqrt{x} + \sqrt{x}\,\sqrt{x + 5} - x}$

$= \dfrac{\sqrt{x + 5} - \sqrt{x}}{5}$

48. $\dfrac{1}{3\sqrt{x} + 1} = \dfrac{1(3\sqrt{x} - 1)}{(3\sqrt{x} + 1)(3\sqrt{x} - 1)} = \dfrac{3\sqrt{x} - 1}{9x - 3\sqrt{x} + 3\sqrt{x} - 1} = \dfrac{3\sqrt{x} - 1}{9x - 1}$

Section 4.7

2. $7 + 2i$, a = 7, b = 2

4. $10 - 7i$, a = 10, b = -7

6. $11 + 2i$, a = 11, b = 2

8. $-1 + \sqrt{3}\,i$, a = -1, b = $\sqrt{3}$

10. $-\frac{1}{5} - \frac{1}{5}i$, $a = -\frac{1}{5}$, $b = -\frac{1}{5}$ 12. $\frac{3}{5} + \frac{2\sqrt{3}}{5}i$, $a = \frac{3}{5}$, $b = \frac{2\sqrt{3}}{5}$

14. $23 + 0i$, $a = 23$, $b = 0$ 16. $0 + 10i$, $a = 0$, $b = 10$

18. $6 + 4i$ 20. $\frac{2 - \sqrt{5}\,i}{2}$ 22. $-6i$ 24. -15

26. T   28. T   30. T

Section 4.8

2. $(8 + 3i) + (5 + 9i) = (8 + 5) + (3 + 9)i = 13 + 12i$

4. $(1 + i) + (7i) = (1 + 0) + (1 + 7)i = 1 + 8i$

6. $(-3 + 4i) + (-3 - 4i) = (-3 - 3) + (4 - 4)i = -6 + 0i = -6$

8. $(15 + 10i) - (4 + 6i) = (15 - 4) + (10 - 6)i = 11 + 4i$

10. $(5 - 4i) - (-2 + 4i) = (5 + 2) + (-4 - 4)i = 7 + (-8i) = 7 - 8i$

12. $(-6 - i) - (-2 - i) = (-6 + 2) + (-1 + 1)i = -4 + 0i = -4$

14. $(\frac{1 + \sqrt{5}\,i}{2}) - (\frac{1 - \sqrt{5}\,i}{2}) = (\frac{1}{2} - \frac{1}{2}) + (\frac{\sqrt{5}}{2} + \frac{\sqrt{5}}{2})i = \frac{2\sqrt{5}\,i}{2} = \sqrt{5}\,i$

16. $i^{28} = (i^4)^7 = 1^7 = 1$

18. $i^{13} = i^{12}i = (i^4)^3 i = 1^3 \cdot i = i$

20. $i^{27} = i^{24}i^3 = (i^4)^6 i^3 = 1^6 \cdot (-i) = -i$

22. $i^{-10} = i^{-12}i^2 = (i^4)^{-3}i^2 = 1^{-3} \cdot (-1) = -1$

24. $(-5i)(2i) = (-5 \cdot 2)i^2 = -10(-1) = 10$

26. $(4 - 3i)(5 + 7i) = 20 + 28i - 15i - 21i^2$

$= 20 + 28i - 15i - 21(-1)$

$= 41 + 13i$

28. $(10 + 9i)^2 = 100 + 180i + 81i^2 = 100 + 180i + 81(-1) = 19 + 180i$

30. $(1 - i)^3$

$= (1 - i)^2(1 - i)$

$= (1 - 2i + i^2)(1 - i)$

$= (-2i)(1 - i)$

$= -2i + 2i^2$

$= -2i - 2$

$= -2 - 2i$

32. $\dfrac{3 + 5i}{1 - i}$

$= \dfrac{(3 + 5i)(1 + i)}{(1 - i)(1 + i)}$

$= \dfrac{3 + 3i + 5i + 5i^2}{1 + i - i - i^2}$

$= \dfrac{3 + 8i + 5(-1)}{1 - (-1)}$

$= \dfrac{-2 + 8i}{2}$

$= -1 + 4i$

34. $\dfrac{1 + 3i}{2i}$

$= \dfrac{(1 + 3i)(-2i)}{(2i)(-2i)}$

$= \dfrac{-2i - 6i^2}{-4i^2}$

$= \dfrac{-2i + 6}{4}$

$= \dfrac{3 - i}{2}$

36. $\dfrac{2}{5 + 3i}$

$= \dfrac{2(5 - 3i)}{(5 + 3i)(5 - 3i)}$

$= \dfrac{2(5 - 3i)}{25 - 15i + 15i - 9i^2}$

$= \dfrac{2(5 - 3i)}{34}$

$= \dfrac{5 - 3i}{17}$

$= \dfrac{5}{17} - \dfrac{3}{17}i$

38. $\dfrac{6i}{3 - 2i}$

$= \dfrac{6i(3 + 2i)}{(3 - 2i)(3 + 2i)}$

$= \dfrac{18i + 12i^2}{9 + 6i - 6i - 4i^2}$

$= \dfrac{18i - 12}{13}$

$= -\dfrac{12}{13} + \dfrac{18}{13}i$

40. $\dfrac{a + bi}{c + di}$

$= \dfrac{(a + bi)(c - di)}{(c + di)(c - di)}$

$= \dfrac{ac - adi + bci - bdi^2}{c^2 - cdi + cdi - d^2i^2}$

$= \dfrac{(ac + bd) - (ad - bc)i}{c^2 + d^2}$

$= \dfrac{ac + bd}{c^2 + d^2} - \dfrac{ad - bc}{c^2 + d^2}i$

CHAPTER FIVE
FIRST-DEGREE
EQUATIONS AND INEQUALITIES

Section 5.1

2. $x + 5 = 9$

     $x + 5 - 5 = 9 - 5$

          $x = 4$

4. $x - 3 = 8$

     $x - 3 + 3 = 8 + 3$

          $x = 11$

6. $5x = 10$

$$\frac{5x}{5} = \frac{10}{5}$$

   $x = 2$

8. $-4x = 32$

$$\frac{-4x}{-4} = \frac{32}{-4}$$

   $x = -8$

10. $3x - 6 = 12$

    $3x - 6 + 6 = 12 + 6$

        $3x = 18$

$$\frac{3x}{3} = \frac{18}{3}$$

      $x = 6$

12. $2x + 7 = 8$

    $2x + 7 - 7 = 8 - 7$

        $2x = 1$

$$\frac{2x}{2} = \frac{1}{2}$$

      $x = \frac{1}{2}$

14. $-x + 13 = 24$

    $-x + 13 - 13 = 24 - 13$

         $-x = 11$

   $(-1) \cdot (-x) = (-1) \cdot 11$

         $x = -11$

16. $8x - 15 = -15$

   $8x - 15 + 15 = -15 + 15$

         $8x = 0$

    $\frac{1}{8} \cdot 8x = \frac{1}{8} \cdot 0$

        $x = 0$

18. $-9x + 11 = 13 - 20$

   $-9x + 11 = -7$

     $-9x = -18$

       $x = 2$

20. $2x + 3x + 1 = 21$

     $5x + 1 = 21$

       $5x = 20$

       $x = 4$

22. $7x - 20x + 5 = 17$

$\quad\ -13x + 5 = 17$

$\qquad\ -13x = 12$

$\qquad\quad x = -\dfrac{12}{13}$

24. $3x + 5 = 2x + 12$

$\quad\ 3x = 2x + 7$

$\quad\ 3x - 2x = 2x - 2x + 7$

$\qquad\quad x = 7$

26. $\quad\ - x + 10 = x + 2$

$\quad\ - x + x + 10 = x + x + 2$

$\qquad\quad 10 = 2x + 2$

$\qquad\quad\ 8 = 2x$

$\qquad\quad\ 4 = x$

28. $3x - 1 = 8x - 1$

$\quad\ 3x = 8x$

$\quad\ 3x - 3x = 8x - 3x$

$\qquad\quad 0 = 5x$

$\qquad\quad 0 = x$

30. $5(x - 1) = 3x - 9$

$\quad\ 5x - 5 = 3x - 9$

$\qquad\ 5x = 3x - 4$

$\qquad\ 2x = - 4$

$\qquad\ x = - 2$

32. $5 - (x - 1) = 2x + 15$

$\quad\ 5 - x + 1 = 2x + 15$

$\qquad\ - x + 6 = 2x + 15$

$\qquad\quad - x = 2x + 9$

$\qquad\quad - 3x = 9$

$\qquad\qquad x = - 3$

34. $6 + 4(2x + 5) = 10 + 12$

$\quad\ 6 + 8x + 20 = 22$

$\qquad\ 8x + 26 = 22$

$\qquad\qquad 8x = - 4$

$\qquad\qquad x = -\dfrac{4}{8}$

$\qquad\qquad x = -\dfrac{1}{2}$

36. $4(x - 5) - 2(6x - 7) = 1$

$\quad\ 4x - 20 - 12x + 14 = 1$

$\qquad\qquad - 8x - 6 = 1$

$\qquad\qquad - 8x = 7$

$\qquad\qquad x = -\dfrac{7}{8}$

38. $x - (10x + 1) = 7 - 5(3x + 4)$

$\quad x - 10x - 1 = 7 - 15x - 20$

$\quad -9x - 1 = -15x - 13$

$\quad -9x = -15x - 12$

$\quad 6x = -12$

$\quad x = -2$

40. $\quad x - (x + 3)^2 = -x^2 - 19$

$\quad x - (x^2 + 6x + 9) = -x^2 - 19$

$\quad x - x^2 - 6x - 9 = -x^2 - 19$

$\quad -x^2 - 5x - 9 = -x^2 - 19$

$\quad -5x - 9 = -19$

$\quad -5x = -10$

$\quad x = 2$

42. When $a = c$, since in that case the denominator is zero.

Section 5.2

2. $\quad \dfrac{x}{3} + \dfrac{x}{6} = 5$

$\quad 6(\dfrac{x}{3} + \dfrac{x}{6}) = 6 \cdot 5$

$\quad 2x + x = 30$

$\quad 3x = 30$

$\quad x = 10$

4. $\quad \dfrac{x}{2} - \dfrac{x}{10} = -2$

$\quad 10(\dfrac{x}{2} - \dfrac{x}{10}) = 10(-2)$

$\quad 5x - x = -20$

$\quad 4x = -20$

$\quad x = -5$

6. $\quad \dfrac{y}{4} + 2 = \dfrac{y}{3} + \dfrac{7}{12}$

$\quad 12(\dfrac{y}{4} + 2) = 12(\dfrac{y}{3} + \dfrac{7}{12})$

$\quad 3y + 24 = 4y + 7$

$\quad 17 = y$

8. $\quad 6.2y - 1.1 = 8.9y + 7$

$\quad 10(6.2y - 1.1) = 10(8.9y + 7)$

$\quad 62y - 11 = 89y + 70$

$\quad -81 = 27y$

$\quad -3 = y$

10. $\quad \dfrac{x + 1}{3} - \dfrac{x}{4} = \dfrac{1}{2}$

$\quad 12(\dfrac{x + 1}{3} - \dfrac{x}{4}) = 12 \cdot \dfrac{1}{2}$

$\quad 4(x + 1) - 3x = 6$

$\quad 4x + 4 - 3x = 6$

$\quad x + 4 = 6$

$\quad x = 2$

12. $\quad \dfrac{a + 5}{4} - \dfrac{3}{8} = \dfrac{3}{4} - \dfrac{a + 8}{16}$

$\quad 16(\dfrac{a + 5}{4} - \dfrac{3}{8}) = 16(\dfrac{3}{4} - \dfrac{a + 8}{16})$

$\quad 4(a + 5) - 6 = 12 - (a + 8)$

$\quad 4a + 20 - 6 = 12 - a - 8$

$\quad 4a + 14 = -a + 4$

$\quad 5a = -10$

$\quad a = -2$

14. $\dfrac{3}{b} + \dfrac{12}{5} = 3$

$5b\left(\dfrac{3}{b} + \dfrac{12}{5}\right) = 5b \cdot 3$

$15 + 12b = 15b$

$15 = 3b$

$5 = b$

16. $\dfrac{1}{t} = \dfrac{1}{2} - \dfrac{1}{3}$

$6t \cdot \dfrac{1}{t} = 6t\left(\dfrac{1}{2} - \dfrac{1}{3}\right)$

$6 = 3t - 2t$

$6 = t$

18. $2 - \dfrac{4}{3m} = \dfrac{2}{m} - \dfrac{4}{3}$

$3m\left(2 - \dfrac{4}{3m}\right) = 3m\left(\dfrac{2}{m} - \dfrac{4}{3}\right)$

$6m - 4 = 6 - 4m$

$10m = 10$

$m = 1$

20. $\dfrac{7x - 2}{x - 2} = 0$

$(x - 2)\,\dfrac{7x - 2}{x - 2} = (x - 2) \cdot 0$

$7x - 2 = 0$

$7x = 2$

$x = \dfrac{2}{7}$

22. $\dfrac{2x}{x + 8} = \dfrac{10}{x + 8}$

$(x + 8)\,\dfrac{2x}{x + 8} = (x + 8)\,\dfrac{10}{x + 8}$

$2x = 10$

$x = 5$

24. $\dfrac{5}{x + 2} = \dfrac{3}{x - 2}$

$(x^2 - 4)\,\dfrac{5}{x + 2} = (x^2 - 4)\,\dfrac{3}{x - 2}$

$(x - 2)5 = (x + 2)3$

$5x - 10 = 3x + 6$

$2x = 16$

$x = 8$

26. $\dfrac{2x}{x - 4} = \dfrac{2x}{x + 4}$

$(x^2 - 16)\,\dfrac{2x}{x - 4} = (x^2 - 16)\,\dfrac{2x}{x + 4}$

$(x + 4)\,2x = (x - 4)\,2x$

$2x^2 + 8x = 2x^2 - 8x$

$8x = -8x$

$16x = 0$

$x = 0$

28.
$$\frac{x}{x - 7} + 1 = \frac{-9}{x - 7}$$

$$(x - 7)(\frac{x}{x - 7} + 1) = (x - 7)\frac{-9}{x - 7}$$

$$x + (x - 7) = -9$$

$$2x - 7 = -9$$

$$2x = -2$$

$$x = -1$$

30.
$$\frac{3}{x} + \frac{2}{x - 1} = \frac{4}{x}$$

$$x(x - 1)(\frac{3}{x} + \frac{2}{x - 1}) = x(x - 1)\frac{4}{x}$$

$$(x - 1)3 + 2x = (x - 1)4$$

$$3x - 3 + 2x = 4x - 4$$

$$5x - 3 = 4x - 4$$

$$x = -1$$

32.
$$\frac{x + 1}{3x + 15} + \frac{x}{x + 5} = \frac{5}{3}$$

$$(3x + 15)(\frac{x + 1}{3x + 15} + \frac{x}{x + 5}) = (3x + 15)\frac{5}{3}$$

$$(x + 1) + 3x = (x + 5)5$$

$$4x + 1 = 5x + 25$$

$$-24 = x$$

34.
$$\frac{4}{x + 2} - \frac{5}{x - 2} = \frac{x}{x^2 - 4}$$

$$(x^2 - 4)(\frac{4}{x + 2} - \frac{5}{x - 2}) = (x^2 - 4)\frac{x}{x^2 - 4}$$

$$(x - 2)4 - (x + 2)5 = x$$

$$4x - 8 - 5x - 10 = x$$

$$-x - 18 = x$$

$$-18 = 2x$$

$$-9 = x$$

36.
$$\frac{x}{x-2} + 4 = \frac{2}{x-2}$$

$$(x-2)\left(\frac{x}{x-2} + 4\right) = (x-2)\frac{2}{x-2}$$

no solution

$$x + (x-2)4 = 2$$

$$x + 4x - 8 = 2$$

$$5x = 10$$

$$x = 2$$

But x = 2 causes two denominators in the original equation to equal zero. Thus the solution set is $\emptyset$.

38.
$$\frac{2}{x^2 + x - 12} = \frac{1}{x+4} + \frac{1}{x-3}$$

$$(x^2 + x - 12)\frac{2}{x^2 + x - 12} = (x^2 + x - 12)\left(\frac{1}{x+4} + \frac{1}{x-3}\right)$$

$$2 = (x-3) + (x+4)$$

$$2 = 2x + 1$$

$$1 = 2x$$

$$\frac{1}{2} = x$$

40.
$$0.15x + 0.45(5 - x) = 3.15$$

$$100[0.15x + 0.45(5-x)] = 100(3.15)$$

$$15x + 45(5-x) = 315$$

$$15x + 225 - 45x = 315$$

$$-30x = 90$$

$$x = -3$$

Section 5.3

2.
$$d = rt$$
$$50 = r\left(\frac{40}{60}\right)$$
$$50 = r\left(\frac{2}{3}\right)$$
$$\frac{3}{2} \cdot 50 = r$$
$$r = 75 \text{ km/hr}$$

4.
$$A = \frac{1}{2}h(b_1 + b_2)$$
$$32 = \frac{1}{2} \cdot 4(10 + b_2)$$
$$32 = 2(10 + b_2)$$
$$16 = 10 + b_2$$
$$b_2 = 6 \text{ in.}$$

6. $C = \frac{5}{9}(F - 32)$

     $100 = \frac{5}{9}(F - 32)$

     $900 = 5(F - 32)$

     $180 = F - 32$

       $F = 212^\circ$

8. $V = \ell wh$

     $1848 = 21 \cdot 11h$

     $1848 = 231h$

       $h = 8$ cm

10. $A = \pi r^2$

     $50.24 = 3.14 r^2$

      $16 = r^2$

       $r = 4$ in.

12. $\frac{1}{S} = \frac{1}{E} - \frac{1}{P}$

    $\frac{1}{P} = \frac{1}{E} - \frac{1}{S}$

    $\frac{1}{P} = \frac{S - E}{SE}$

    $P = \frac{SE}{S - E}$

    $P = \frac{(779.95)(365.26)}{779.95 - 365.26}$

    $P = 687$ days

14. $\frac{1}{R} = \frac{1}{R_1} + \frac{1}{R_2}$

       $\frac{1}{9} = \frac{1}{12} + \frac{1}{R_2}$

  $36R_2 \, \frac{1}{9} = 36R_2 \left(\frac{1}{12} + \frac{1}{R_2} \right)$

      $4R_2 = 3R_2 + 36$

      $R_2 = 36$ ohms

16. $A = P + Pr$

     $A - P = Pr$

     $\frac{A - P}{P} = r$

18. $d = rt$

    $\frac{d}{r} = \frac{rt}{r}$

    $\frac{d}{r} = t$

20. $A = \frac{1}{3}Bh$

     $3A = Bh$

     $\frac{3A}{B} = h$

22. $V = \ell wh$

    $\frac{V}{\ell h} = w$

24. $5y - x + 10 = 0$

         $5y = x - 10$

          $y = \frac{x - 10}{5}$

26.
$$y = \frac{1}{x - 1}$$

$$y(x - 1) = 1$$

$$yx - y = 1$$

$$yx = 1 + y$$

$$x = \frac{1 + y}{y}$$

28.
$$h = -16t^2 + vt$$

$$h + 16t^2 = vt$$

$$\frac{h + 16t^2}{t} = v$$

30.
$$A = \frac{1}{2}h(b_1 + b_2)$$

$$2A = h(b_1 + b_2)$$

$$\frac{2A}{h} = b_1 + b_2$$

$$\frac{2A}{h} - b_1 = b_2$$

32.
$$4x - 3y + 1 = 0$$

$$4x + 1 = 3y$$

$$\frac{4x + 1}{3} = y$$

34.
$$S = \frac{a - rt}{1 - r}$$

$$S(1 - r) = a - rt$$

$$S - Sr = a - rt$$

$$rt - Sr = a - S$$

$$r(t - S) = a - S$$

$$r = \frac{a - S}{t - S}$$

36.
$$\frac{1}{R} = \frac{1}{R_1} + \frac{1}{R_2}$$

$$RR_1R_2 \frac{1}{R} = RR_1R_2 \left(\frac{1}{R_1} + \frac{1}{R_2}\right)$$

$$R_1R_2 = RR_2 + RR_1$$

$$R_1R_2 - RR_2 = RR_1$$

$$R_2(R_1 - R) = RR_1$$

$$R_2 = \frac{RR_1}{R_1 - R}$$

Section 5.4

2. x = smaller integer

$x + 2$ = larger integer

$x + x + 2 = 176$

$$2x = 174$$

$$x = 87$$

$$x + 2 = 89$$

4. x = shortest length

$2x$ = middle length

$3x$ = longest length

$x + 2x + 3x = 24$

$$6x = 24$$

$$x = 4 \text{ ft}$$

$$2x = 8 \text{ ft}$$

$$3x = 12 \text{ ft}$$

6.   w = width

   3w + 2 = length

   w + 3w + 2 = 22

   4w = 20

   w = 5m

   3w + 2 = 17m

8.   w = width

   w + 3 = length

   $(w + 1)(w + 4) = w(w + 3) + 14$

   $w^2 + 5w + 4 = w^2 + 3w + 14$

   5w + 4 = 3w + 14

   2w = 10

   w = 5m

   w + 3 = 8m

10.   x = Pam's age

   x + 11 = Pat's age

   x + 11 + 3 = 2(x + 3)

   x + 14 = 2x + 6

   - x = - 8

   x = 8 yr

   x + 11 = 19 yr

12.   x = amount of 15% alloy

   25 - x = amount of 40% alloy

   .15x + .40(25 - x) = .20(25)

   15x + 40(25 - x) = 20(25)

   15x + 1000 - 40x = 500

   - 25x = - 500

   x = 20g

   25 - x = 5 g

14. x = amount of formaldehyde

   x + .40(15) = .50(x + 15)

   10x + 4(15) = 5(x + 15)

   10x + 60 = 5x + 75

   5x = 15

   $x = 3 \text{ cm}^3$

16.　　x = no. of nickels

x + 3 = no. of dimes

2x = no. of quarters

5x + 10(x + 3) + 25(2x) = 745

5x + 10x + 30 + 50x = 745

65x = 715

x = 11 nickels

x + 3 = 14 dimes

2x = 22 quarters

18. t = time to overtake 1st jogger

distance of 2nd jogger = distance of 1st jogger

$$12t = 10(t + \frac{1}{4})$$

$$12t = 10t + \frac{5}{2}$$

$$2t = \frac{5}{2}$$

$$t = \frac{5}{4} = 1\frac{1}{4} \text{ hr}$$

20. r = rate of current

4 mi upstream time = 6 mi downstream time

$$\frac{4}{10 - r} = \frac{6}{10 + r}$$

$$(100 - r^2)\frac{4}{10 - r} = (100 - r^2)\frac{6}{10 + r}$$

$$(10 + r)4 = (10 - r)6$$

$$40 + 4r = 60 - 6r$$

$$10r = 20$$

$$r = 2 \text{ mph}$$

22. t = time for both pipes together

$$\frac{1}{8} + \frac{1}{6} = \frac{1}{t}$$

$$24t\left(\frac{1}{8} + \frac{1}{6}\right) = 24t\,\frac{1}{t}$$

$$3t + 4t = 24$$

$$7t = 24$$

$$t = \frac{24}{7} = 3\,\frac{3}{7}\ hr$$

24. t = time for new computer

$$\frac{1}{3} + \frac{1}{t} = \frac{1}{\frac{4}{3}}$$

$$\frac{1}{3} + \frac{1}{t} = \frac{3}{4}$$

$$12t\left(\frac{1}{3} + \frac{1}{t}\right) = 12t\,\frac{3}{4}$$

$$4t + 12 = 9t$$

$$-5t = -12$$

$$t = \frac{12}{5} = 2\,\frac{2}{5}\ hr$$

26. x = amount of bonus

$$x - .20x = 500$$

$$.8x = 500$$

$$8x = 5000$$

$$x = \$625$$

28. h = no. of work hours

$$8h - .30(8h) = 280$$

$$8h - 2.4h = 280$$

$$5.6h = 280$$

$$h = 50\ hr$$

30. x = amount invested at 7%

$$.06(5000) + .07x = 510$$

$$300 + .07x = 510$$

$$.07x = 210$$

$$7x = 21000$$

$$x = \$3000$$

32.   x = no. of calculators

$$C = 17x + 1050$$

$$R = 19.5x$$

$$R = C$$

$$19.5x = 17x + 1050$$

$$2.5x = 1050$$

$$x = 420\ calculators$$

Section 5.5

2. $7x < 21$

    $\frac{1}{7} \cdot 7x < \frac{1}{7} \cdot 21$

       $x < 3$

4. $-4x > 20$

    $(-\frac{1}{4})(-4x) < (-\frac{1}{4})20$

        $x < -5$

6. $4x - 9 < 3$

    $4x - 9 + 9 < 3 + 9$

        $4x < 12$

         $x < 3$

8. $2x + 5 > 13$

    $2x + 5 - 5 > 13 - 5$

        $2x > 8$

         $x > 4$

10. $-3x - 4 \leq 2$

      $-3x \leq 6$

  $(-\frac{1}{3})(-3x) \geq (-\frac{1}{3})6$

       $x \geq -2$

12. $-x + 17 \geq 7$

      $-x \geq -10$

  $(-1)(-x) \leq (-1)(-10)$

       $x \leq 10$

14. $6x - 8 < x + 2$

     $6x < x + 10$

     $5x < 10$

      $x < 2$

16. $10 - 3x \leq 7x - 10$

      $-3x \leq 7x - 20$

     $-10x \leq -20$

      $x \geq 2$

18. $5x + 1 \geq 2x + 1$

     $3x \geq 0$

  $\frac{1}{3} \cdot 3x \geq \frac{1}{3} \cdot 0$

     $x \geq 0$

20. $4(x - 3) > 3x - 5$

    $4x - 12 > 3x - 5$

      $4x > 3x + 7$

       $x > 7$

22. $2(x + 2) - 5(2x - 4) < 0$

    $2x + 4 - 10x + 20 < 0$

       $-8x + 24 < 0$

        $-8x < -24$

         $x > 3$

24. $3x - (15x + 14) < x + 4$

    $3x - 5x - 14 < x + 4$

      $-2x - 14 < x + 4$

        $-3x < 18$

        $x > -6$

26. $6(2x - 1) \leq 6 - 3(x + 4)$

    $12x - 6 \leq 6 - 3x - 12$

    $12x - 6 \leq -3x - 6$

         $15x \leq 0$

           $x \leq 0$

28. $\dfrac{x}{3} - 2 \geq x + \dfrac{2}{3}$

    $3(\dfrac{x}{3} - 2) \geq 3(x + \dfrac{2}{3})$

      $x - 6 \geq 3x + 2$

        $-2x \geq 8$

          $x \leq -4$

30. $\dfrac{1}{9}x + \dfrac{5}{18} > \dfrac{1}{6}x + \dfrac{2}{3}$

    $18(\dfrac{1}{9}x + \dfrac{5}{18}) > 18(\dfrac{1}{6}x + \dfrac{2}{3})$

      $2x + 5 > 3x + 12$

           $-x > 7$

           $x < -7$

32. $1 < 10 - \dfrac{4x + 5}{5}$

      $5 < 50 - (4x + 5)$

      $5 < 50 - 4x - 5$

      $5 < -4x + 45$

      $4x < 40$

       $x < 10$

34. $3 < x + 4 < 8$

    $3 - 4 < x + 4 - 4 < 8 - 4$

     $-1 < x < 4$

36. $-4 \leq 2x \leq 7$

    $-\dfrac{4}{2} \leq \dfrac{2x}{2} \leq \dfrac{7}{2}$

    $-2 \leq x \leq \dfrac{7}{2}$

38. $-10 \leq 5x - 10 < 0$

    $0 \leq 5x < 10$

    $0 \leq x < 2$

40. $-1 < \dfrac{-8x - 3}{3} \leq 1$

    $-3 < -8x - 3 \leq 3$

     $0 < -8x \leq 6$

      $0 > x \geq -\dfrac{3}{4}$

    $-\dfrac{3}{4} \leq x < 0$

42. $C = \dfrac{5}{9}(F - 32)$

    $-10 < C < 25$

    $-10 < \dfrac{5}{9}(F - 32) < 25$

    $-18 < F - 32 < 45$

    $14^\circ < F < 77^\circ$

44.   x = remaining wins

    $\dfrac{86 + x}{125 + 37} \geq .65$

    $\dfrac{86 + x}{162} \geq .65$

    $86 + x \geq 105.3$

        $x \geq 19.3$

    or $x \geq 20$ games

46. x = no. of magazines

C = .50x + 1000

R = .40x + .25(x - 5000)

= .40x + .25x - 1250

= .65x - 1250

P = R - C

= (.65x - 1250) - (.50x + 1000)

= .15x - 2250

$$P > 0$$

.15x - 2250 > 0

.15x > 2250

x > 15,000 magazines

48. p = price of calculator

12 + .30(12) $\leq$ p < 24

12 + 3.60 $\leq$ p < 24

$15.60 $\leq$ p < $24

Section 5.6

2. $|y| = 8$

y = 8 or y = - 8

4. $|y + 3| = 2$

y + 3 = 2 or y + 3 = - 2

y = - 1 or y = - 5

6. $|3y - 1| = 10$

3y - 1 = 10 or 3y - 1 = - 10

y $= \frac{11}{3}$ or y = - 3

8. $|y - \frac{3}{4}| = \frac{1}{4}$

y $- \frac{3}{4} = \frac{1}{4}$ or y $- \frac{3}{4} = -\frac{1}{4}$

y = 1 or y $= \frac{1}{2}$

10. $|2y - 8| = 0$

2y - 8 = 0

y = 4

12. $|5y| - 7 = 8$

    $|5y| = 15$

    $5y = 15$ or $5y = -15$

    $y = 3$ or $y = -3$

14. $|y| < 2$

    $-2 < y < 2$

16. $|y - 3| < 4$

    $-4 < y - 3 < 4$

    $-1 < y < 7$

18. $|2y + 5| \leq 3$

    $-3 \leq 2y + 5 \leq 3$

    $-8 \leq 2y \leq -2$

    $-4 \leq y \leq -1$

20. $|y - 5| < 0.01$

    $-0.01 < y - 5 < 0.01$

    $4.99 < y < 5.01$

22. $|y| > 3$

    $y > 3$ or $y < -3$

24. $|y - 2| > 4$

    $y - 2 > 4$ or $y - 2 < -4$

    $y > 6$ or $y < -2$

26. $|4 - 3y| \geq 10$

    $4 - 3y \geq 10$ or $4 - 3y \leq -10$

    $-3y \geq 6$ or $-3y \leq -14$

    $y \leq -2$ or $y \geq \dfrac{14}{3}$

28. $|a| = a$ if and only if $a \geq 0$. Therefore

$|x - 2| = x - 2$ if and only if $x - 2 \geq 0$. That is,

when $x \geq 2$.

30. $|5 + (-2)| = |3| = 3$ but $|5| + |-2| = 5 + 2 = 7$.

32. $|ay + b| \geq c$

    $ay + b \geq c$ or $ay + b \leq -c$

        $ay \geq c - b$ or $ay \leq -c - b$

        $y \geq \dfrac{c - b}{a}$ or $y \leq -\dfrac{c + b}{a}$

CHAPTER SIX
SECOND-DEGREE EQUATIONS AND INEQUALITIES

Section 6.1

2. $4x^2 + 5x = 9$

     $4x^2 + 5x - 9 = 0$

     $a = 4, \; b = 5, \; c = -9$

4. $x^2 = x + \sqrt{3}$

     $x^2 - x - \sqrt{3} = 0$

     $a = 1, \; b = -1, \; c = -\sqrt{3}$

6. $8x^2 = 1$

     $8x^2 - 1 = 0$

     $a = 8, \; b = 0, \; c = -1$

8. $2(5x^2 - 3) = 3x - 6$

     $10x^2 - 6 = 3x - 6$

     $10x^2 - 3x = 0$

     $a = 10, \; b = -3, \; c = 0$

10. $x(x + 8) = 8x$

     $x^2 + 8x = 8x$

        $x^2 = 0$

     $a = 1, \; b = 0, \; c = 0$

12. $x^2 - 5x + 6 = 0$

     $(x - 2)(x - 3) = 0$

     $x - 2 = 0 \quad x - 3 = 0$

       $x = 2 \qquad x = 3$

14. $x^2 + 8x + 15 = 0$

     $(x + 3)(x + 5) = 0$

     $x + 3 = 0 \quad x + 5 = 0$

       $x = -3 \qquad x = -5$

16. $x^2 + x = 12$

     $x^2 + x - 12 = 0$

     $(x+4)(x-3) = 0$

     $x + 4 = 0 \quad x - 3 = 0$

       $x = -4 \qquad x = 3$

18. $x^2 - 6x - 16 = 0$

$(x - 8)(x + 2) = 0$

$x - 8 = 0 \quad x + 2 = 0$

$x = 8 \qquad x = -2$

20. $x^2 + 9 = 6x$

$x^2 - 6x + 9 = 0$

$(x - 3)(x - 3) = 0$

$x - 3 = 0 \quad x - 3 = 0$

$x = 3$

22. $2x^2 + 9x - 5 = 0$

$(2x - 1)(x + 5) = 0$

$2x - 1 = 0 \quad x + 5 = 0$

$x = \dfrac{1}{2} \qquad x = -5$

24. $4x^2 - 11x + 6 = 0$

$(4x - 3)(x - 2) = 0$

$4x - 3 = 0 \quad x - 2 = 0$

$x = \dfrac{3}{4} \qquad x = 2$

26. $5x^2 = 3 - 2x$

$5x^2 + 2x - 3 = 0$

$(5x - 3)(x + 1) = 0$

$5x - 3 = 0 \quad x + 1 = 0$

$x = \dfrac{3}{5} \qquad x = -1$

28. $36x^2 + 12x + 1 = 0$

$(6x + 1)(6x + 1) = 0$

$6x + 1 = 0$

$x = -\dfrac{1}{6}$

30. $x^2 + 7x = 0$

$x(x + 7) = 0$

$x = 0 \quad x + 7 = 0$

$x = -7$

32. $10x^2 = 2x$

$10x^2 - 2x = 0$

$2x(5x - 1) = 0$

$2x = 0 \quad 5x - 1 = 0$

$x = 0 \qquad x = \dfrac{1}{5}$

34. $2(x^2 - 4) = (x - 2)^2$

$2x^2 - 8 = x^2 - 4x + 4$

$x^2 + 4x - 12 = 0$

$(x + 6)(x - 2) = 0$

$x + 6 = 0 \quad x - 2 = 0$

$x = -6 \qquad x = 2$

36. $x(3x + 8) = (x + 2)(x + 5)$

$3x^2 + 8x = x^2 + 7x + 10$

$2x^2 + x - 10 = 0$

$(2x + 5)(x - 2) = 0$

$2x + 5 = 0 \qquad x - 2 = 0$

$x = -\dfrac{5}{2} \qquad x = 2$

38. $10x^3 + 30x^2 + 20x = 0$

$10x(x^2 + 3x + 2) = 0$

$10x(x + 1)(x + 2) = 0$

$10x = 0 \quad x + 1 = 0 \quad x + 2 = 0$

$x = 0 \qquad x = -1 \qquad x = -2$

40. $2x^4 - 7x^3 - 4x^2 = 0$

$x^2(2x^2 - 7x - 4) = 0$

$x^2(2x + 1)(x - 4) = 0$

$x^2 = 0 \quad 2x + 1 = 0 \quad x - 4 = 0$

$x = 0 \qquad x = -\dfrac{1}{2} \qquad x = 4$

Section 6.2

2. $x^2 - 9 = 0$

$(x - 3)(x + 3) = 0$

$x - 3 = 0 \quad x + 3 = 0$

$x = 3 \qquad x = -3$

4. $x^2 - 100 = 0$

$(x - 10)(x + 10) = 0$

$x - 10 = 0 \quad x + 10 = 0$

$x = 10 \qquad x = -10$

6. $16x^2 = 1$

$x^2 = \dfrac{1}{16}$

$x = \pm\dfrac{1}{4}$

8. $4x^2 - 81 = 0$

$4x^2 = 81$

$x^2 = \dfrac{81}{4}$

$x = \pm\dfrac{9}{2}$

10. $x^2 = 2$

$x = \pm\sqrt{2}$

12. $x^2 - 20 = 0$

$x^2 = 20$

$x = \pm\sqrt{20}$

$x = \pm 2\sqrt{5}$

14. $x^2 + 9 = 0$

$x^2 = -9$

$x = \pm\sqrt{-9}$

$x = \pm 3i$

16. $x^2 + 7 = 0$

$x^2 = -7$

$x = \pm\sqrt{-7}$

$x = \pm\sqrt{7}\ i$

18. $\qquad 36x^2 + 121 = 0$

$(6x + 11i)(6x - 11i) = 0$

$6x + 11i = 0 \qquad 6x - 11i = 0$

$6x = -11i \qquad 6x = 11i$

$x = -\dfrac{11}{6}i \qquad x = \dfrac{11}{6}i$

20. $\qquad (x - 1)^2 = 25 \quad$ or $\quad (x - 1)^2 = 25$

$(x - 1)^2 - 5^2 = 0 \qquad\qquad x - 1 = \pm 5$

$[(x - 1) - 5][(x - 1) + 5] = 0 \qquad\qquad x = 1 \pm 5$

$(x - 6)(x + 4) = 0 \qquad x = 1 + 5 \quad x = 1 - 5$

$x - 6 = 0 \quad x + 4 = 0 \qquad x = 6 \qquad x = -4$

$x = 6 \qquad x = -4$

22. $(x + 4)^2 = 3$

$x + 4 = \pm\sqrt{3}$

$x = -4 \pm\sqrt{3}$

24. $(x - \dfrac{1}{2})^2 = \dfrac{25}{4}$

$x - \dfrac{1}{2} = \pm\dfrac{5}{2}$

$x = \dfrac{1}{2} \pm \dfrac{5}{2}$

$x = 3 \qquad x = -2$

26. $(x + \frac{1}{2})^2 = \frac{11}{4}$

$\qquad x + \frac{1}{2} = \pm \frac{\sqrt{11}}{2}$

$\qquad\qquad x = -\frac{1}{2} \pm \frac{\sqrt{11}}{2}$

$\qquad\qquad x = \frac{-1 \pm \sqrt{11}}{2}$

28. $\frac{x^2}{4} - \frac{7x}{4} + 3 \quad = 0$

$\qquad x^2 - 7x + 12 \quad = 0$

$\qquad (x - 3)(x - 4) = 0$

$\qquad x - 3 = 0 \quad x - 4 = 0$

$\qquad\qquad x = 3 \qquad x = 4$

30. $\frac{x^2}{4} - \frac{5x}{6} - \frac{2}{3} \quad = 0$

$\qquad 3x^2 - 10x - 8 \quad = 0$

$\qquad (3x + 2)(x - 4) = 0$

$\qquad 3x + 2 = 0 \quad x - 4 = 0$

$\qquad\qquad x = -\frac{2}{3} \qquad x = 4$

32. $\frac{x^2}{5} - \frac{x}{6} \quad = 0$

$\qquad 6x^2 - 5x \quad = 0$

$\qquad x(6x - 5) \quad = 0$

$\qquad x = 0 \quad 6x - 5 = 0$

$\qquad\qquad x = \frac{5}{6}$

34. $1 + \frac{2}{x} - \frac{8}{x^2} \quad = 0$

$\quad x^2 + 2x - 8 \quad = 0$

$\quad (x + 4)(x - 2) = 0$

$\quad x + 4 = 0 \quad x - 2 = 0$

$\qquad x = -4 \qquad x = 2$

36. $\qquad\qquad \frac{3}{2x} - \frac{2}{2x + 1} - 2 \quad = 0$

$\quad 3(2x + 1) - 2(2x) - 2(2x)(2x + 1) = 0$

$\qquad\qquad 6x + 3 - 4x - 8x^2 - 4x = 0$

$\qquad\qquad\qquad -8x^2 - 2x + 3 = 0$

$\qquad (-1) \cdot (-8x^2 - 2x + 3) = (-1) \cdot 0$

$\qquad\qquad\qquad 8x^2 + 2x - 3 = 0$

$\qquad\qquad (4x + 3)(2x - 1) = 0$

$\qquad 4x + 3 = 0 \quad 2x - 1 = 0$

$\qquad\qquad x = -\frac{3}{4} \qquad x = \frac{1}{2}$

38. $\dfrac{x}{x + 2} + \dfrac{4}{x - 1} = 2$

$x(x - 1) + 4(x + 2) = 2(x + 2)(x - 1)$

$x^2 - x + 4x + 8 = 2x^2 + 2x - 4$

$-x^2 + x + 12 = 0$

$x^2 - x - 12 = 0$

$(x - 4)(x + 3) = 0$

$x - 4 = 0 \quad x + 3 = 0$

$x = 4 \qquad x = -3$

40. $1 = \dfrac{1}{x + 3} + \dfrac{20}{(x + 3)^2}$

$(x + 3)^2 = (x + 3) + 20$

$x^2 + 6x + 9 = x + 23$

$x^2 + 5x - 14 = 0$

$(x + 7)(x - 2) = 0$

$x + 7 = 0 \quad x - 2 = 0$

$x = -7 \qquad x = 2$

Section 6.3

2. $x^2 - 6x + 4 = 0$

$x^2 - 6x = -4$

$x^2 - 6x + 9 = -4 + 9$

$(x - 3)^2 = 5$

$x - 3 = \pm \sqrt{5}$

$x = 3 \pm \sqrt{5}$

4. $x^2 - 4x - 3 = 0$

$x^2 - 4x = 3$

$x^2 - 4x + 4 = 3 + 4$

$(x - 2)^2 = 7$

$x - 2 = \pm \sqrt{7}$

$x = 2 \pm \sqrt{7}$

6. $x^2 + 8x - 2 = 0$

$x^2 + 8x = 2$

$x^2 + 8x + 16 = 2 + 16$

$(x + 4)^2 = 18$

$x + 4 = \pm\sqrt{18}$

$x = -4 \pm 3\sqrt{2}$

8. $x^2 + x - 3 = 0$

$x^2 + x = 3$

$x^2 + x + \frac{1}{4} = 3 + \frac{1}{4}$

$(x + \frac{1}{2})^2 = \frac{13}{4}$

$x + \frac{1}{2} = \pm\frac{\sqrt{13}}{2}$

$x = -\frac{1}{2} \pm \frac{\sqrt{13}}{2}$

10. $x^2 - 4x + 5 = 0$

$x^2 - 4x = -5$

$x^2 - 4x + 4 = -5 + 4$

$(x - 2)^2 = -1$

$x - 2 = \pm\sqrt{-1}$

$x - 2 = \pm i$

$x = 2 \pm i$

12. $x^2 + 3x - 9 = 0$

$x^2 + 3x = 9$

$x^2 + 3x + \frac{9}{4} = 9 + \frac{9}{4}$

$(x + \frac{3}{2})^2 = \frac{45}{4}$

$x + \frac{3}{2} = \pm\frac{3\sqrt{5}}{2}$

$x = -\frac{3}{2} \pm \frac{3\sqrt{5}}{2}$

14. $4x^2 - 4x - 3 = 0$

$x^2 - x - \frac{3}{4} = 0$

$x^2 - x = \frac{3}{4}$

$x^2 - x + \frac{1}{4} = \frac{3}{4} + \frac{1}{4}$

$(x - \frac{1}{2})^2 = 1$

$x - \frac{1}{2} = \pm 1$

$x = \frac{1}{2} \pm 1$

$x = \frac{1}{2} + 1 \quad x = \frac{1}{2} - 1$

$x = \frac{3}{2} \qquad x = -\frac{1}{2}$

16. $2x^2 + x - 2 = 0$

$x^2 + \frac{1}{2}x - 1 = 0$

$x^2 + \frac{1}{2}x = 1$

$x^2 + \frac{1}{2}x + \frac{1}{16} = 1 + \frac{1}{16}$

$(x + \frac{1}{4})^2 = \frac{17}{16}$

$x + \frac{1}{4} = \pm\frac{\sqrt{17}}{4}$

$x = -\frac{1}{4} \pm \frac{\sqrt{17}}{4}$

18. $5x^2 + 10x - 4 = 0$

$x^2 + 2x - \dfrac{4}{5} = 0$

$x^2 + 2x \qquad = \dfrac{4}{5}$

$x^2 + 2x + 1 = \dfrac{4}{5} + 1$

$(x + 1)^2 = \dfrac{9}{5}$

$x + 1 = \pm \dfrac{3}{\sqrt{5}}$

$x + 1 = \pm \dfrac{3\sqrt{5}}{5}$

$x = -1 \pm \dfrac{3\sqrt{5}}{5}$

20. $4x^2 - 12x + 57 = 0$

$x^2 - 3x + \dfrac{57}{4} = 0$

$x^2 - 3x \qquad = -\dfrac{57}{4}$

$x^2 - 3x + \dfrac{9}{4} = -\dfrac{57}{4} + \dfrac{9}{4}$

$\left(x - \dfrac{3}{2}\right)^2 = -\dfrac{48}{4}$

$\left(x - \dfrac{3}{2}\right)^2 = -16$

$x - \dfrac{3}{2} = \pm \sqrt{-16}$

$x - \dfrac{3}{2} = \pm 4i$

$x = \dfrac{3}{2} \pm 4i$

Section 6.4

2. $x^2 + 2x - 3 = 0$

a) $(x + 3)(x - 1) = 0$

$x = -3 \qquad x = 1$

b) $a = 1 \quad b = 2 \quad c = -3$

$x = \dfrac{-2 \pm \sqrt{2^2 - 4(1)(-3)}}{2(1)}$

$x = \dfrac{-2 \pm \sqrt{4 - (-12)}}{2}$

$x = \dfrac{-2 \pm \sqrt{16}}{2}$

$x = \dfrac{-2 + 4}{2} \qquad x = \dfrac{-2 - 4}{2}$

$x = 1 \qquad\qquad x = -3$

4. $x^2 - 4x - 21 = 0$

a) $(x - 7)(x + 3) = 0$

   $x = 7 \quad x = -3$

b) $a = 1 \quad b = -4 \quad c = -21$

$$x = \frac{-(-4) \pm \sqrt{(-4)^2 - 4(1)(-21)}}{2(1)}$$

$$x = \frac{4 \pm \sqrt{16 - (-84)}}{2}$$

$$x = \frac{4 \pm \sqrt{100}}{2}$$

$$x = \frac{4 + 10}{2} \qquad x = \frac{4 - 10}{2}$$

$$x = 7 \qquad\qquad x = -3$$

6. $x^2 - 4x + 4 = 0$

a) $(x - 2)^2 = 0$

   $x = 2$

b) $a = 1 \quad b = -4 \quad c = 4$

$$x = \frac{-(-4) \pm \sqrt{(-4)^2 - 4(1)(4)}}{2(1)}$$

$$x = \frac{4 \pm \sqrt{16 - 16}}{2}$$

$$x = \frac{4 \pm \sqrt{0}}{2}$$

$$x = \frac{4 + 0}{2} \qquad x = \frac{4 - 0}{2}$$

$$x = 2$$

8. $9x^2 - 1 = 0$

   a) $(3x - 1)(3x + 1) = 0$

       $x = \dfrac{1}{3}$ $x = -\dfrac{1}{3}$

   b) $a = 9$ $b = 0$ $c = -1$

       $x = \dfrac{-0 \pm \sqrt{0^2 - 4(9)(-1)}}{2(9)}$

       $x = \dfrac{0 \pm \sqrt{0 - (-36)}}{18}$

       $x = \dfrac{\pm \sqrt{36}}{18}$

       $x = \dfrac{+6}{18}$ $x = \dfrac{-6}{18}$

       $x = \dfrac{1}{3}$ $x = -\dfrac{1}{3}$

10. $x^2 + 6x = 0$

   a) $x(x + 6) = 0$

       $x = 0$ $x = -6$

   b) $a = 1$ $b = 6$ $c = 0$

       $x = \dfrac{-6 \pm \sqrt{6^2 - 4(1)(0)}}{2(1)}$

       $x = \dfrac{-6 \pm \sqrt{36 - 0}}{2}$

       $x = \dfrac{-6 + 6}{2}$ $x = \dfrac{-6 - 6}{2}$

       $x = 0$ $x = -6$

12. $x^2 + x - 3 = 0$

      $x = \dfrac{-1 \pm \sqrt{1^2 - 4(1)(-3)}}{2(1)}$

      $x = \dfrac{-1 \pm \sqrt{1 + 12}}{2}$

      $x = \dfrac{-1 \pm \sqrt{13}}{2}$

14. $x^2 - 5x + 5 = 0$

$$x = \frac{-(-5) \pm \sqrt{(-5)^2 - 4(1)(5)}}{2(1)}$$

$$x = \frac{5 \pm \sqrt{25 - 20}}{2}$$

$$x = \frac{5 \pm \sqrt{5}}{2}$$

16. $x^2 - 4x + 2 = 0$

$$x = \frac{-(-4) \pm \sqrt{(-4)^2 - 4(1)(2)}}{2(1)}$$

$$x = \frac{4 \pm \sqrt{16 - 8}}{2}$$

$$x = \frac{4 \pm \sqrt{8}}{2}$$

$$x = \frac{4 \pm 2\sqrt{2}}{2}$$

$$x = 2 \pm \sqrt{2}$$

18. $x^2 - x + 2 = 0$

$$x = \frac{-(-1) \pm \sqrt{(-1)^2 - 4(1)(2)}}{2(1)}$$

$$x = \frac{1 \pm \sqrt{1 - 8}}{2}$$

$$x = \frac{1 \pm \sqrt{-7}}{2}$$

$$x = \frac{1 \pm \sqrt{7}\,i}{2}$$

20. $3x^2 + 4x - 5 = 0$

$$x = \frac{-4 \pm \sqrt{4^2 - 4(3)(-5)}}{2(3)}$$

$$x = \frac{-4 \pm \sqrt{16 + 60}}{6}$$

$$x = \frac{-4 \pm \sqrt{76}}{6}$$

$$x = \frac{-4 \pm 2\sqrt{19}}{6}$$

$$x = \frac{-2 \pm \sqrt{19}}{3}$$

22. $5x^2 + 2x + 1 = 0$

$$x = \frac{-2 \pm \sqrt{2^2 - 4(5)(1)}}{2(5)}$$

$$x = \frac{-2 \pm \sqrt{4 - 20}}{10}$$

$$x = \frac{-2 \pm \sqrt{-16}}{10}$$

$$x = \frac{-2 \pm 4i}{10}$$

$$x = \frac{-1 \pm 2i}{5}$$

24. $4x^2 - 11x + 6 = 0$

$$x = \frac{-(-11) \pm \sqrt{(-11)^2 - 4(4)(6)}}{2(4)}$$

$$x = \frac{11 \pm \sqrt{121 - 96}}{8}$$

$$x = \frac{11 \pm \sqrt{25}}{8}$$

$$x = \frac{11 + 5}{8} \qquad x = \frac{11 - 5}{8}$$

$$x = 2 \qquad\qquad x = \frac{3}{4}$$

26. $\dfrac{x^2}{6} - \dfrac{x}{3} - 1 = 0$

$$x^2 - 2x - 6 = 0$$

$$x = \frac{-(-2) \pm \sqrt{(-2)^2 - 4(1)(-6)}}{2(1)}$$

$$x = \frac{2 \pm \sqrt{4 + 24}}{2}$$

$$x = \frac{2 \pm \sqrt{28}}{2}$$

$$x = \frac{2 \pm 2\sqrt{7}}{2}$$

$$x = 1 \pm \sqrt{7}$$

28. $\dfrac{3x^2}{2} - x - \dfrac{1}{3} = 0$

$$9x^2 - 6x - 2 = 0$$

$$x = \frac{-(-6) \pm \sqrt{(-6)^2 - 4(9)(-2)}}{2(9)}$$

$$x = \frac{6 \pm \sqrt{36 + 72}}{18}$$

$$x = \frac{6 \pm \sqrt{108}}{18}$$

$$x = \frac{6 \pm 6\sqrt{3}}{18}$$

$$x = \frac{1 \pm \sqrt{3}}{3}$$

30. $1 - \dfrac{\sqrt{6}}{x} - \dfrac{1}{x^2} = 0$

$$x^2 - \sqrt{6}\, x - 1 = 0$$

$$x = \frac{-(-\sqrt{6}) \pm \sqrt{(-\sqrt{6})^2 - 4(1)(-1)}}{2(1)}$$

$$x = \frac{\sqrt{6} \pm \sqrt{6 + 4}}{2}$$

$$x = \frac{\sqrt{6} \pm \sqrt{10}}{2}$$

32. $b^2 - 4ac = (-4)^2 - 4(2)(3) = -8$. Two imaginary solutions.

34. $b^2 - 4ac = 1^2 - 4(1)(-3) = 13$.   Two real solutions.

36. $b^2 - 4ac = (-10)^2 - 4(5\sqrt{5})(\sqrt{5}) = 0$. One real solution.

38. $\qquad\quad b^2 - 4ac = 0$

$(-k)^2 - 4(2)(2) \quad = 0$

$\qquad\quad k^2 - 16 = 0$

$\qquad\qquad\quad k^2 = 16$

$\qquad\qquad\quad k = \pm 4$

40. $r_1 \cdot r_2 = \dfrac{-b + \sqrt{b^2 - 4ac}}{2a} \cdot \dfrac{-b - \sqrt{b^2 - 4ac}}{2a}$

$\qquad\quad = \dfrac{b^2 + b\sqrt{b^2 - 4ac} - b\sqrt{b^2 - 4ac} - (b^2 - 4ac)}{4a^2}$

$\qquad\quad = \dfrac{4ac}{4a^2}$

$\qquad\quad = \dfrac{c}{a}$

Section 6.5

2. $\qquad$ x = first number

$\quad$ x + 1 = second number

$\quad$ x + 2 = third number

$x^2 + (x + 1)^2 + (x + 2)^2 \qquad = 149$

$x^2 + x^2 + 2x + 1 + x^2 + 4x + 4 = 149$

$\qquad\qquad\quad 3x^2 + 6x + 5 = 149$

$\qquad\qquad\quad 3x^2 + 6x - 144 = 0$

$\qquad\quad (3x + 24)(x - 6) = 0$

$\qquad\qquad\qquad \cancel{x = -8} \qquad\quad x = 6$

$\qquad\qquad\qquad\qquad\qquad\quad x + 1 = 7$

$\qquad\qquad\qquad\qquad\qquad\quad x + 2 = 8$

4. x = the integer

$6x^2 - 3x \quad\quad = 84$

$6x^2 - 3x - 84 = 0$

$2x^2 - x - 28 \ = 0$

$(2x + 7)(x - 4) = 0$

$\cancel{x = -\dfrac{7}{2}} \quad\quad x = 4$

6. x = the number

$\dfrac{1}{x}$ = the reciprocal

$x + \dfrac{1}{x} \quad\quad\quad = \dfrac{25}{12}$

$12x^2 + 12 \quad\quad = 25x$

$12x^2 - 25x + 12 \ = 0$

$(4x - 3)(3x - 4) = 0$

$x = \dfrac{3}{4} \quad\quad x = \dfrac{4}{3}$

$\dfrac{1}{x} = \dfrac{4}{3} \quad\quad \dfrac{1}{x} = \dfrac{3}{4}$

8. $\quad\quad w$ = width

$3w - 3$ = length

$w(3w - 3) \quad\quad = 90$

$3w^2 - 3w \quad\quad = 90$

$w^2 - w \quad\quad\quad = 30$

$w^2 - w - 30 \quad = 0$

$(w + 5)(w - 6) = 0$

$\cancel{w = -5} \quad\quad w = 6 \text{ m}$

$\quad\quad\quad 3w - 3 = 15 \text{ m}$

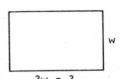

w

3w - 3

10. x = width of walk

$4x^2 + 2(4x) + 2(6x) = 4\cdot 6$

$4x^2 + 8x + 12x \quad\quad = 24$

$4x^2 + 20x - 24 \quad\quad = 0$

$x^2 + 5x - 6 \quad\quad\quad = 0$

$(x + 6)(x - 1) \quad\quad = 0$

$\cancel{x = -6} \quad\quad x = 1 \text{ ft}$

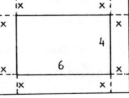

12. x = length of wire

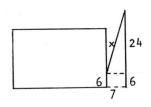

$$x^2 = 24^2 + 7^2$$

$$x^2 = 576 + 49$$

$$x^2 = 625$$

$$x \cancel{=} -25 \qquad x = 25 \text{ ft}$$

14. x = length of side

$$x^2 + x^2 = 72^2$$

$$2x^2 = 5184$$

$$x^2 = 2592$$

$$x \cancel{=} -\sqrt{2592}$$

$$x = \sqrt{2592}$$

$$x = \sqrt{1296 \cdot 2}$$

$$x = 36\sqrt{2} \text{ cm}$$

16. r = rate of current

time downstream + time upstream = $2\frac{1}{2}$

$$\frac{12}{10 + r} + \frac{12}{10 - r} = \frac{5}{2}$$

$$12(10 - r) + 12(10 + r) = \frac{5}{2}(100 - r^2)$$

$$120 - 12r + 120 + 12r = 250 - \frac{5}{2}r^2$$

$$\frac{5}{2}r^2 = 10$$

$$r^2 = 4$$

$$r \cancel{=} -2 \qquad r = 2 \text{ mph}$$

18. t = time for first pipe

   $t - 2$ = time for second pipe

   $\dfrac{1}{t} + \dfrac{1}{t-2} = \dfrac{1}{4}$

   $4(t-2) + 4t = t(t-2)$

   $4t - 8 + 4t = t^2 - 2t$

   $-t^2 + 10t - 8 = 0$

   $t^2 - 10t + 8 = 0$

   $t = \dfrac{-(-10) \pm \sqrt{(-10)^2 - 4(1)(8)}}{2(1)}$

   $t = \dfrac{10 \pm \sqrt{100 - 32}}{2}$

   $t = \dfrac{10 \pm \sqrt{68}}{2}$

   $t = \dfrac{10 \pm 2\sqrt{17}}{2}$

   ~~$t = 5 - \sqrt{17}$~~ $t = 5 + \sqrt{17} \approx 9.12$ hr

                    $t - 2 = 3 + \sqrt{17} \approx 7.12$ hr

20. $h = -16t^2 + 96t$

   (a) $h = 0$ (b) $h = 128$

   $-16t^2 + 96t = 0$ $-16t^2 + 96t = 128$

   $t^2 - 6t = 0$ $t^2 - 6t = -8$

   $t(t - 6) = 0$ $t^2 - 6t + 8 = 0$

   $t = 0$ sec $t = 6$ sec $(t - 2)(t - 4) = 0$

                    $t = 2$ sec $t = 4$ sec

(c)
$$h = 160$$

$$-16t^2 + 96t = 160$$

$$t^2 - 6t = -10$$

$$t^2 - 6t + 10 = 0$$

$$t = \frac{-(-6) \pm \sqrt{(-6)^2 - 4(1)(10)}}{2(1)}$$

$$t = \frac{6 \pm \sqrt{36 - 40}}{2}$$

$$t = \frac{6 \pm \sqrt{-4}}{2}$$

$$t = \frac{6 \pm 2i}{2}$$

$$t = 3 \pm i$$

Since these are imaginary numbers, we conclude that the pellet never reaches a height of 160 ft.

22. $x = 700 - 4p$

$$4p = 700 - x$$

$$p = 175 - \frac{1}{4}x$$

$$R = px = (175 - \frac{1}{4}x)x = 175x - \frac{1}{4}x^2$$

$$C = 20x$$

Set $R = C$ to find break-even points.

$$175x - \frac{1}{4}x^2 = 20x$$

$$700x - x^2 = 80x$$

$$0 = x^2 - 620x$$

$$0 = x(x - 620)$$

$$x = 0 \text{ items} \qquad x = 620 \text{ items}$$

24. $x = 60 - \frac{1}{2} p$

$\frac{1}{2} p = 60 - x$

$p = 120 - 2x$

$R = px = (120 - 2x)x = 120x - 2x^2$

$C = 25x + 600$

Set R = C to find break-even points.

$120x - 2x^2 = 25x + 600$

$0 = 2x^2 - 95x + 600$

$0 = (2x - 15)(x - 40)$

$x = \frac{15}{2} = 7.5 \qquad x = 40$

Since x is in hundreds, the break-even points occur at 750 items and 4000 items.

Section 6.6

2. $\sqrt{x} = 5$

$(\sqrt{x})^2 = 5^2$

$x = 25$

4. $\sqrt{x - 1} = 4$

$(\sqrt{x - 1})^2 = 4^2$

$x - 1 = 16$

$x = 17$

6. $\sqrt{3x + 1} + 2 = 0$

$\sqrt{3x + 1} = -2$

$(\sqrt{3x + 1})^2 = (-2)^2$

$3x + 1 = 4$

$3x = 3$

$x = 1$

But x = 1 does not check in the original equation. Therefore, the solution set is ∅.

8. $\sqrt{x - 2} = 4 - x$

$(\sqrt{x - 2})^2 = (4 - x)^2$

$x - 2 = 16 - 8x + x^2$

$0 = x^2 - 9x + 18$

$0 = (x - 3)(x - 6)$

$x = 3 \qquad x = 6$

Since only x = 3 checks in the original equation, the solution set is {3}.

10. $\sqrt{x + 7} - \sqrt{x} = 1$

$\sqrt{x + 7} = 1 + \sqrt{x}$

$(\sqrt{x + 7})^2 = (1 + \sqrt{x})^2$

$x + 7 = 1 + 2\sqrt{x} + x$

$6 = 2\sqrt{x}$

$3 = \sqrt{x}$

$3^2 = (\sqrt{x})^2$

$9 = x$

12. $\sqrt{3x + 4} - \sqrt{x} = 2$

$\sqrt{3x + 4} = 2 + \sqrt{x}$

$(\sqrt{3x + 4})^2 = (2 + \sqrt{x})^2$

$3x + 4 = 4 + 4\sqrt{x} + x$

$2x = 4\sqrt{x}$

$x = 2\sqrt{x}$

$x^2 = (2\sqrt{x})^2$

$x^2 = 4x$

$x^2 - 4x = 0$

$x(x - 4) = 0$

$x = 0 \qquad x = 4$

14. $\sqrt{5x + 6} - \sqrt{x + 3} = 3$

$\sqrt{5x + 6} = 3 + \sqrt{x + 3}$

$5x + 6 = 9 + 6\sqrt{x + 3} + (x + 3)$

$4x - 6 = 6\sqrt{x + 3}$

$2x - 3 = 3\sqrt{x + 3}$

$4x^2 - 12x + 9 = 9(x + 3)$

$4x^2 - 21x - 18 = 0$

$(4x + 3)(x - 6) = 0$

$x = -\dfrac{3}{4} \qquad x = 6$

Since $x = -\dfrac{3}{4}$ does not check, the solution set is just $\{6\}$.

16. $x^{\frac{2}{3}} - x^{\frac{1}{3}} - 6 = 0$ 18. $x^{\frac{1}{2}} - 5x^{\frac{1}{4}} + 6 = 0$

$(x^{\frac{1}{3}} - 3)(x^{\frac{1}{3}} + 2) = 0$ $(x^{\frac{1}{4}} - 2)(x^{\frac{1}{4}} - 3) = 0$

$x^{\frac{1}{3}} - 3 = 0 \quad x^{\frac{1}{3}} + 2 = 0$ $x^{\frac{1}{4}} - 2 = 0 \quad x^{\frac{1}{4}} - 3 = 0$

$x^{\frac{1}{3}} = 3 \qquad x^{\frac{1}{3}} = -2$ $x^{\frac{1}{4}} = 2 \qquad x^{\frac{1}{4}} = 3$

$(x^{\frac{1}{3}})^3 = 3^3 \quad (x^{\frac{1}{3}})^3 = (-2)^3$ $(x^{\frac{1}{4}})^4 = 2^4 \quad (x^{\frac{1}{4}})^4 = 3^4$

$\quad x = 27 \qquad\qquad x = -8$ $\quad x = 16 \qquad\qquad x = 81$

20. $3x + 5x^{\frac{1}{2}} - 2 = 0$ 22. $x^4 - 10x^2 + 9 = 0$

$(3x^{\frac{1}{2}} - 1)(x^{\frac{1}{2}} + 2) = 0$ $(x^2 - 1)(x^2 - 9) = 0$

$3x^{\frac{1}{2}} - 1 = 0 \quad x^{\frac{1}{2}} + 2 = 0$ $x^2 - 1 = 0 \quad x^2 - 9 = 0$

$x^{\frac{1}{2}} = \frac{1}{3} \qquad x^{\frac{1}{2}} = -2$ $x^2 = 1 \qquad\qquad x^2 = 9$

$x = \frac{1}{9} \qquad\qquad x = 4$ $x = \pm 1 \qquad\qquad x = \pm 3$

Since $x = 4$ does not check,
the solution set is just
$\{\frac{1}{9}\}$.

24. $x^4 - 7x^2 + 12 = 0$

$(x^2 - 4)(x^2 - 3) = 0$

$x^2 - 4 = 0 \quad x^2 - 3 = 0$

$x^2 = 4 \qquad\qquad x^2 = 3$

$x = \pm 2 \qquad\qquad x = \pm\sqrt{3}$

26. $(x^2 + 2)^2 - 19(x^2 + 2) + 18 = 0$

$[(x^2 + 2) - 1][(x^2 + 2) - 18] = 0$

$[x^2 + 1][x^2 - 16] = 0$

$x^2 + 1 = 0 \qquad x^2 - 16 = 0$

$x^2 = -1 \qquad x^2 = 16$

$x = \pm i \qquad x = \pm 4$

28. $A = \pi r^2$

$\dfrac{A}{\pi} = r^2$

$\pm\sqrt{\dfrac{A}{\pi}} = r$

30. $A = P(1 + r)^2$

$\dfrac{A}{P} = (1 + r)^2$

$\pm\sqrt{\dfrac{A}{P}} = 1 + r$

$r = -1 \pm\sqrt{\dfrac{A}{P}}$

32. $x^2 + y^2 = 4$

$y^2 = 4 - x^2$

$y = \pm\sqrt{4 - x^2}$

34. $x^2 - y^2 = 1$

$- y^2 = 1 - x^2$

$y^2 = x^2 - 1$

$y = \pm\sqrt{x^2 - 1}$

36. $t = \sqrt{\dfrac{2s}{g}}$

$t^2 = \dfrac{2s}{g}$

$gt^2 = 2s$

$\dfrac{gt^2}{2} = s$

38. $x^2 - 2yx + y^2 - 1 = 0$

$a = 1 \quad b = -2y \quad c = y^2 - 1$

$x = \dfrac{2y \pm\sqrt{4y^2 - 4(y^2 - 1)}}{2}$

$x = \dfrac{2y \pm\sqrt{4}}{2}$

$x = y \pm 1$

Section 6.7

2. d   4. f   6. c

8. $(x - 3)(x + 1) < 0$

$$\underset{\underset{-1}{}}{\text{no}} \quad \underset{\underset{3}{}}{\text{yes}} \quad \text{no}$$

$$-1 < x < 3$$

10. $x(x - 2) > 0$

$$\text{yes} \quad \underset{\underset{0}{}}{\text{no}} \quad \underset{\underset{2}{}}{\text{yes}}$$

$$x < 0 \quad \text{or} \quad x > 2$$

12. $x^2 - 5x + 4 \leq 0$

$(x - 1)(x - 4) \leq 0$

$$\underset{\underset{1}{}}{\text{no}} \quad \underset{\underset{4}{}}{\text{yes}} \quad \text{no}$$

$$1 \leq x \leq 4$$

14. $3x^2 + 5x < 2$

$3x^2 + 5x - 2 < 0$

$(3x - 1)(x + 2) < 0$

$$\underset{\underset{-2}{}}{\text{no}} \quad \underset{\underset{\frac{1}{3}}{}}{\text{yes}} \quad \text{no}$$

$$-2 < x < \frac{1}{3}$$

16. $x^2 - 1 \geq 0$

$(x - 1)(x + 1) \geq 0$

$$\text{yes} \quad \underset{\underset{-1}{}}{\text{no}} \quad \underset{\underset{1}{}}{\text{yes}}$$

$$x \leq -1 \quad \text{or} \quad x \geq 1$$

18. $x^2 < 5$

$x^2 - 5 < 0$

$(x - \sqrt{5})(x + \sqrt{5}) < 0$

$$\underset{\underset{-\sqrt{5}}{}}{\text{no}} \quad \underset{\underset{\sqrt{5}}{}}{\text{yes}} \quad \text{no}$$

$$-\sqrt{5} < x < \sqrt{5}$$

20. $x^2 + 1 > 0$

Since $x^2 \geq 0$ for every real number x, then $x^2 + 1 > 0$ for every real number x. Thus, the solution set is R.

22. $(x + 2)(x - 2)^2(x - 4) \geq 0$

$$\text{yes} \quad \underset{\underset{-2}{}}{} \quad \text{no} \quad \underset{\underset{2}{}}{} \quad \text{no} \quad \underset{\underset{4}{}}{\text{yes}}$$

$$x \leq -2 \quad \text{or} \quad x = 2 \quad \text{or} \quad x \geq 4$$

24. $x^4 - 9x^2 < 0$

$x^2(x^2 - 9) < 0$

$x^2(x - 3)(x + 3) < 0$

$$\underset{\underset{-3}{}}{\text{no}} \quad \underset{\underset{0}{}}{\text{yes}} \quad \underset{\underset{3}{}}{\text{yes}} \quad \text{no}$$

$$-3 < x < 0 \quad \text{or} \quad 0 < x < 3$$

26. $\dfrac{6}{x - 4} \leq -1$

$$\frac{6}{x - 4} + 1 \leq 0$$

$$\frac{6}{x-4} + \frac{x-4}{x-4} \leq 0$$

$$\frac{x + 2}{x - 4} \leq 0$$

$$\underset{\underset{-2}{}}{\text{no}} \quad \underset{\underset{4}{}}{\text{yes}} \quad \text{no}$$

$$-2 \leq x < 4$$

28. $\dfrac{x}{x + 1} \geq \dfrac{-4}{x}$　　　　30.　　　　　　　$P \geq 16$

$\dfrac{x}{x + 1} + \dfrac{4}{x} \geq 0$　　　　　　　$-4x^2 + 20x \geq 16$

$\dfrac{x \cdot x + 4(x + 1)}{(x + 1)x} \geq 0$　　　　　$-4x^2 + 20x - 16 \geq 0$

$\dfrac{x^2 + 4x + 4}{(x + 1)x} \geq 0$　　　　　　$4x^2 - 20x + 16 \leq 0$

$\dfrac{(x + 2)^2}{(x + 1)x} \geq 0$　　　　　　　$x^2 - 5x + 4 \leq 0$

　　　　　　　　　　　　　$(x - 1)(x - 4) \leq 0$

yes	no	no	yes
	-1	-2	0

no	yes	no
	1	4

$x < -1 \quad \text{or} \quad x = -2 \text{ or } x > 0$　　　　　$1 \leq x \leq 4$

32. $C = 1300x + 6000$　　　　　　　$x = 86 - \dfrac{p}{100}$

$R = px$　　　　　　　　　　$100x = 8600 - p$

$\quad = 8600x - 100x^2$　　　　　　$p = 8600 - 100x$

$P = R - C$

$\quad = (8600x - 100x^2) - (1300x + 6000)$

$\quad = -100x^2 + 7300x - 6000$

　　　　　　　　　$P \geq 15,000$

$-100x^2 + 7300x - 6000 \geq 15,000$

$\quad x^2 - 73x + 60 \leq -150$

$\quad x^2 - 73x + 210 \leq 0$

$\quad (x - 3)(x - 70) \leq 0$

no	yes	no
	3	70

$3 \leq x \leq 70$

CHAPTER SEVEN
FIRST-DEGREE FUNCTIONS AND RELATIONS

Section 7.1

2. Yes	4. Yes	6. No	8. No
10. Yes	12. Yes	14. No	16. Yes
18. Yes	20. No	22. R	

24.
$$y = \frac{1}{x^2 - 16}$$

$$y = \frac{1}{(x - 2)(x + 2)}$$

Domain = $\{x \mid x \neq 2, -2\}$

26.
$$y = \frac{x + 1}{x^2 + 3x}$$

$$y = \frac{x + 1}{x(x + 3)}$$

Domain = $\{x \mid x \neq 0, -3\}$

28.
$$y = \sqrt{3x - 12}$$
$$3x - 12 \geq 0$$
$$x \geq 4$$
Domain = $\{x \mid x \geq 4\}$

30.
$$y = \sqrt{25 - 4x^2}$$
$$25 - 4x^2 \geq 0$$
$$(5 - 2x)(5 + 2x) \geq 0$$

yes   no   yes

$-\frac{5}{2}$ $\frac{5}{2}$

Domain = $\{x \mid x \leq -\frac{5}{2}\} \cup \{x \mid x \geq \frac{5}{2}\}$

32. R

34. $\{y \mid y \geq 0\}$

36. $\{y \mid y \geq -5\}$

38. $\{y \mid y \geq 0\}$

40. $\{y \mid y \leq 3\}$

42. $\{y \mid y \neq 1\}$

44. $f(x) = x + 1$

     $f(0) = 0 + 1 = 1$

     $f(3) = 3 + 1 = 4$

     $f(-2) = -2 + 1 = -1$

     $f(a) = a + 1$

  $f(a + b) = a + b + 1$

46. $g(x) = 3x^2 + 5x - 1$

     $g(0) = 3(0)^2 + 5(0) - 1 = -1$

     $g(3) = 3(3)^2 + 5(3) - 1 = 41$

     $g(-2) = 3(-2)^2 + 5(-2) - 1 = 1$

     $g(a) = 3a^2 + 5a - 1$

  $g(a + b) = 3(a + b)^2 + 5(a + b) - 1$

48. $h(x) = \dfrac{x}{x - 1}$ 50. $f(x) = \dfrac{1}{x + 5}$

     $h(0) = \dfrac{0}{0 - 1} = -1$ $f(x + h) = \dfrac{1}{x + h + 5}$

     $h(3) = \dfrac{3}{3 - 1} = \dfrac{3}{2}$

     $h(-2) = \dfrac{-2}{-2 - 1} = \dfrac{2}{3}$

     $h(a) = \dfrac{a}{a - 1}$

  $h(a + b) = \dfrac{a + b}{a + b - 1}$

52. Since $f(x) = \dfrac{10}{x}$ then $f(-5) = \dfrac{10}{-5} = -2$.

   Since $g(x) = 2x + 5$ then $g(-2) = 2(-2) + 5 = 1$.

   Therefore, $g(f(-5)) = g(-2) = 1$.

54. $g(5 + 1) = g(6) = 6^2 = 36$ but $g(5) + g(1) = 5^2 + 1^2 = 26$.

   Conclusion: $g(a + b) \neq g(a) + g(b)$.

2. Domain = {4, -1, 2}

   Range = {3, 0, -5, 4}

   Not a function

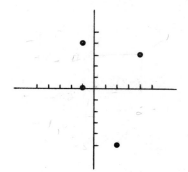

4. Domain = {5, -5, 1, -1}

   Range = {1, 5}

   Is a function

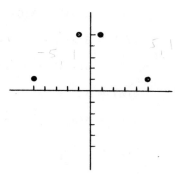

6. Domain = {2, 0}

   Range = {0, 3}

   Not a function

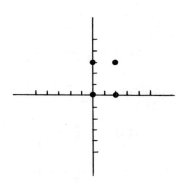

8. No   10. Yes   12. No   14. No

16. Yes

18. I = 2x   x ≥ 0

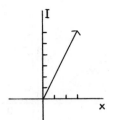

20. C = .5n + 5   n ≥ 0

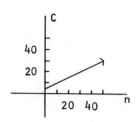

22. T = 2h - 3   0 ≤ h ≤ 5

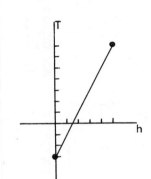

24. T = -3h + 10

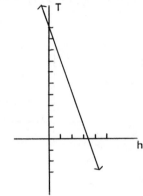

26. T = -4

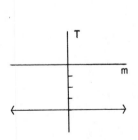

28. $h = -16t^2 + 48t$ 0 ≤ t ≤ 3

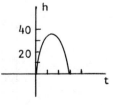

30. $V = \frac{4}{3} \pi r^3 \qquad r \geq 0$

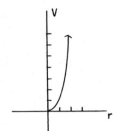

32. $F = \frac{9}{5} C + 32$

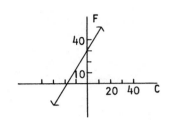

34. Domain = $\{x \mid x \geq 1\}$

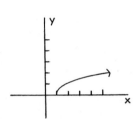

36. Domain = R

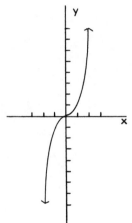

38. Domain = $\{x \mid x \neq 2\}$

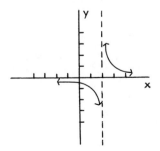

2. $d = \sqrt{(7 - 10)^2 + (4 - 5)^2} = \sqrt{9 + 1} = \sqrt{10}$

   $m = \dfrac{7 - 10}{4 - 5} = \dfrac{-3}{-1} = 3$

4. $d = \sqrt{[2 - (-1)]^2 + [0 - 6]^2} = \sqrt{9 + 36} = \sqrt{45} = 3\sqrt{5}$

   $m = \dfrac{2 - (-1)}{0 - 6} = \dfrac{3}{-6} = -\dfrac{1}{2}$

6. $d = \sqrt{(0 - 1)^2 + (0 - 5)^2} = \sqrt{1 + 25} = \sqrt{26}$

   $m = \dfrac{0 - 1}{0 - 5} = \dfrac{-1}{-5} = \dfrac{1}{5}$

8. $d = \sqrt{[-6 - 3]^2 + [4 - (-4)]^2} = \sqrt{81 + 64} = \sqrt{145}$

   $m = \dfrac{-6 - 3}{4 - (-4)} = \dfrac{-9}{8} = -\dfrac{9}{8}$

10. $d = \sqrt{[\frac{1}{3} - 0]^2 + [-\frac{1}{4} - (-\frac{1}{2})]^2} = \sqrt{\frac{1}{9} + \frac{1}{16}} = \sqrt{\frac{25}{144}} = \dfrac{5}{12}$

    $m = \dfrac{\frac{1}{3} - 0}{-\frac{1}{4} - (-\frac{1}{2})} = \dfrac{\frac{1}{3}}{\frac{1}{4}} = \dfrac{4}{3}$

12.   14.

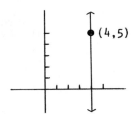

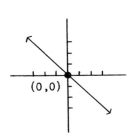

16.

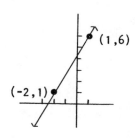

18. $m_1 = \dfrac{-4 - 2}{-1 - 1} = \dfrac{-6}{-2} = 3$

$m_2 = \dfrac{3 - 2}{-3 - 0} = \dfrac{1}{-3} = -\dfrac{1}{3}$

Therefore, $m_1 \cdot m_2 = 3(-\dfrac{1}{3}) = -1$.

20. $d_{AB} = 3 + 5 = 8$

$d_{AC} = 4 + 2 = 6$

$d_{BC} = \sqrt{[4 - (-2)]^2 + [5 - (-3)]^2}$

$= \sqrt{36 + 64}$

$= \sqrt{100}$

$= 10$

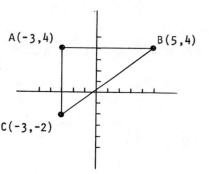

Therefore, the perimeter $= 8 + 6 + 10 = 24$.

22. $m_{AB} = \dfrac{8 - (-3)}{5 - 1} = \dfrac{11}{4}$

$m_{BC} = \dfrac{-3 - (-13)}{1 - (-3)} = \dfrac{10}{4} = \dfrac{5}{2}$

Since segments AB and BC have

different slopes, the points

A, B, and C are not collinear.

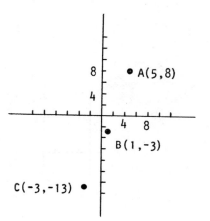

24. $\dfrac{-2 - y}{3 - 1} = \dfrac{1}{4}$

$\dfrac{-2 - y}{2} = \dfrac{1}{4}$

$2(-2 - y) = 1$

$-4 - 2y = 1$

$-2y = 5$

$y = -\dfrac{5}{2}$

26. gradient $= \dfrac{1500 \text{ ft}}{250 \text{ mi}} = 6 \text{ ft/mi}$

Section 7.4

2.

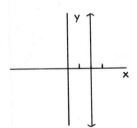

4. $3x + 7 = 0$

$x = -\dfrac{7}{3}$

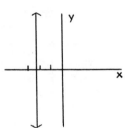

6.

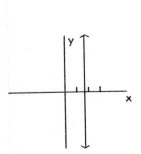

8.

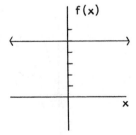

10.

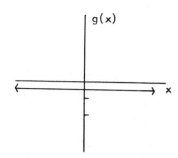

12.

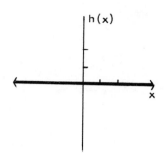

14. $\dfrac{\text{x} \quad | \quad 0 \mid 4}{\text{f(x)} \mid -2 \mid 0}$

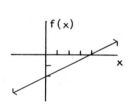

16. $\dfrac{\text{x} \quad | \quad 0 \mid 5}{\text{f(x)} \mid 5 \mid 0}$

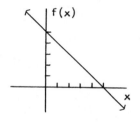

18. $\dfrac{\text{x} \quad | 0 \mid 1}{\text{f(x)} \mid 0 \mid 4}$

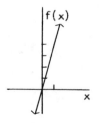

20. $\dfrac{\text{x} \mid 0 \mid 5}{\text{y} \mid 2 \mid 0}$

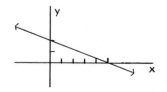

22. $\dfrac{x \;\;\big|\; 0 \;\;\big|\; -7}{y \;\;\big|\; \frac{7}{3} \;\big|\; 0}$

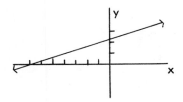

24. $\dfrac{x \;\;\big|\; 0 \;\;\big|\; 2}{y \;\;\big|\; 0 \;\;\big|\; -2}$

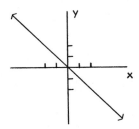

26. $\dfrac{x \;\;\big|\; 0 \;\;\big|\; 3}{y \;\;\big|\; 0 \;\;\big|\; 1}$

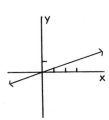

28. $\dfrac{p \;\;\big|\; 0 \;\;\big|\; 100}{x \;\;\big|\; 50 \;\big|\; 0}$

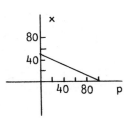

Section 7.5

2. $\quad y - y_1 = m(x - x_1)$

$\quad\quad y - 5 = 3(x - 2)$

$\quad\quad y - 5 = 3x - 6$

$-3x + y + 1 = 0$

or

$\quad 3x - y - 1 = 0$

4. $\quad y - y_1 = m(x - x_1)$

$\quad\quad y - 4 = \dfrac{1}{2}[x - (-3)]$

$\quad\quad 2y - 8 = x + 3$

$-x + 2y - 11 = 0$

or

$\quad x - 2y + 11 = 0$

6. $y - (-4) = -(x - 4)$

    $y + 4 = -x + 4$

    $x + y = 0$

8. $y - 6 = 0(x - 3)$

    $y - 6 = 0$

10. $y - 0 = 6(x - 0)$

    $y = 6x$

    $-6x + y = 0$

or

    $6x - y = 0$

12. $m = \dfrac{-8 - 2}{4 - (-6)} = \dfrac{-10}{10} = -1$

    $y - 2 = -(x + 6)$

    $y - 2 = -x - 6$

    $x + y + 4 = 0$

14. $m = \dfrac{-1 - 1}{-3 - 0} = \dfrac{-2}{-3} = \dfrac{2}{3}$

    $y - 1 = \dfrac{2}{3}(x - 0)$

    $3y - 3 = 2x$

    $-2x + 3y - 3 = 0$

16. $m = \dfrac{8 - 8}{8 - 4} = \dfrac{0}{4} = 0$

Thus, the line is horizontal
and its equation is $y = 8$,
or $y - 8 = 0$.

18. $m = \dfrac{3 - 5}{-2 - (-2)} = \dfrac{-2}{0}$ undefined

Thus, the line is vertical and
its equation is $x = -2$,
or $x + 2 = 0$.

20. $y = mx + b$

    $y = 2x + 5$

    $-2x + y - 5 = 0$

22. $y = \dfrac{1}{3}x + (-1)$

    $3y = x - 3$

    $-x + 3y + 3 = 0$

24. $y = -\dfrac{2}{5}x + 2$

    $5y = -2x + 10$

    $2x + 5y - 10 = 0$

26. $y = 0 \cdot x + \left(-\dfrac{1}{15}\right)$

    $y = -\dfrac{1}{15}$

    $15y = -1$

    $15y + 1 = 0$

28. Since the line passes
through $(0, 6)$, we know $b = 6$.

    $y = \sqrt{3}x + 6$

    $-\sqrt{3}x + y - 6 = 0$

30. $4x + y = 5$

$y = -4x + 5$

$m = -4$

32. $3x + 4y = 12$

$4y = -3x + 12$

$y = -\dfrac{3}{4}x + 3$

$m = -\dfrac{3}{4}$

34. $6x - y = -1$

$-y = -6x - 1$

$y = 6x + 1$

$m = 6$

36. $x - 10y = 2$

$-10y = -x + 2$

$y = \dfrac{1}{10}x - \dfrac{1}{5}$

$m = \dfrac{1}{10}$

38. $y + 2 = 0$

$y = 0 \cdot x - 2$

$m = 0$

40. $x - 5 = 0$

$x = 5$

This is a vertical line and its slope is undefined.

42. $x + y = 5 \qquad\qquad y = x$

$y = -x + 5 \qquad m_2 = 1$

$m_1 = -1$

Since $m_1 \cdot m_2 = -1$, the lines are perpendicular.

44. $4x - 2y = 1$

$-2y = -4x + 1$

$y = 2x - \dfrac{1}{2}$

$m = 2$

Thus, the parallel line through (3, -4) has equation

$y + 4 = 2(x - 3)$

$y + 4 = 2x - 6$

$-2x + y + 10 = 0$

46. $A = (1, 5) \qquad B = (-2, 4)$

$m_{AB} = \dfrac{5 - 4}{1 - (-2)} = \dfrac{1}{3}$

Thus, the perpendicular line through (0, 4) has equation

$y - 4 = -3(x - 0)$

$y - 4 = -3x$

$3x + y - 4 = 0$

48. $(p_1, x_1) = (25, 500)$ $(p_2, x_2) = (35, 450)$

$$m = \frac{500 - 450}{25 - 35} = \frac{50}{-10} = -5$$

$$x - 500 = -5(p - 25)$$

$$x - 500 = -5p + 125$$

$$x = 625 - 5p$$

Section 7.6

2.

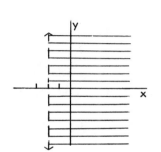

4.

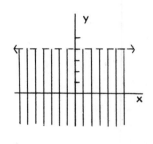

6.

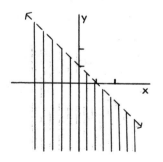

8.

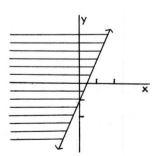

10.

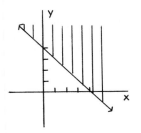

12.

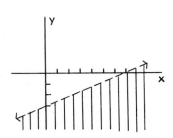

14.

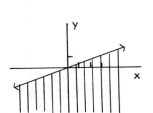

16.

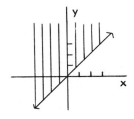

18. $|x| > 3$

   $x > 3$ or $x < -3$

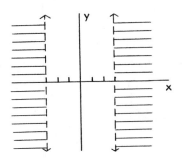

20. $|y| < 2$

   $-2 < y < 2$

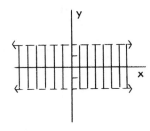

22. $|y| \neq 1$

   $y \neq 1$ and $y \neq -1$

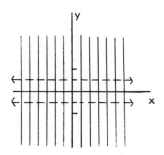

24. $xy \leq 0$

   $(x \geq 0$ and $y \leq 0)$ or $(x \leq 0$ and $y \geq 0)$

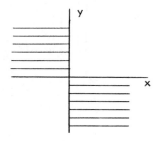

26.

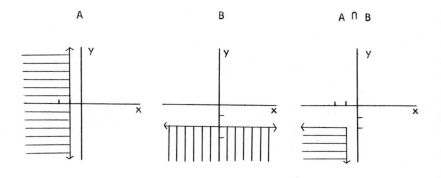

28.

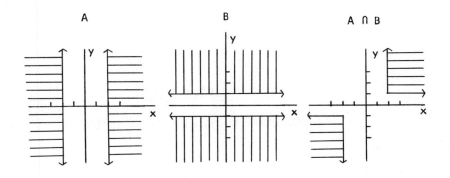

A   B   A ∩ B

30. $4A + 8B \leq 80$

$\quad A \geq 0$

$\quad B \geq 0$

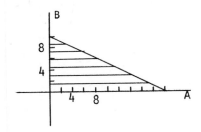

CHAPTER EIGHT
SECOND-DEGREE FUNCTIONS AND RELATIONS

Section 8.1

2. Domain = R

   Range = $\{y \mid y \geq -4\}$

4. Domain = R

   Range = $\{y \mid y \geq -4\}$

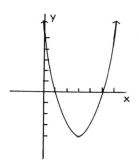

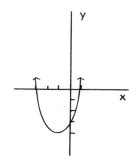

6. Domain = R

   Range = $\{y \mid y \leq \frac{25}{4}\}$

8. Domain = R

   Range = $\{y \mid y \geq 0\}$

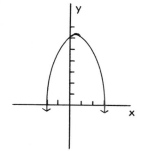

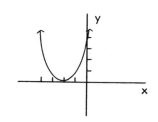

10. Domain = R

Range = $\{y \mid y \geq 1\}$

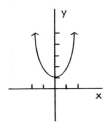

12. Domain = R

Range = $\{y \mid y \leq 5\}$

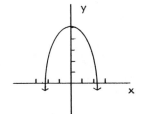

14. Domain = R

Range = $\{y \mid y \leq 0\}$

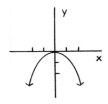

16. Domain = R

Range = $\{y \mid y \geq -\frac{25}{4}\}$

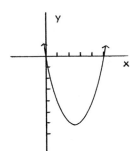

18. Domain = R

Range = $\{y \mid y \geq \frac{1}{2}\}$

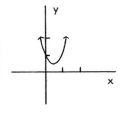

20. Not a function. In this case y is squared instead of x.

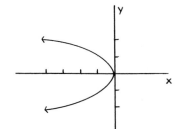

22.

x = width of pasture

$A = x(12 - 2x)$

$A = -2x^2 + 12x$

Max occurs at vertex.

$x = \dfrac{-b}{2a} = \dfrac{-12}{2(-2)} = 3$ mi

$12 - 2x = 6$ mi

24. $h = -16t^2 + 128t$

Max occurs at vertex.

$t = \dfrac{-b}{2a} = \dfrac{-128}{2(-16)} = 4$ sec

$h = -16(4)^2 + 128(4) = 256$ ft

26. $C = 20x + 1000$

$R = px$

$\quad = (80 - \frac{1}{2}x)x$

$\quad = 80x - \frac{1}{2}x^2$

$P = R - C$

$\quad = (80x - \frac{1}{2}x^2) - (20x + 1000)$

$\quad = -\frac{1}{2}x^2 + 60x - 1000$

Max profit occurs at vertex.

$x = \dfrac{-b}{2a} = \dfrac{-60}{2(-\frac{1}{2})} = 60$ units

$p = 80 - \frac{1}{2}(60) = \$50$

$x = 160 - 2p$

$2p = 160 - x$

$p = 80 - \frac{1}{2}x$

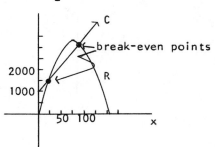

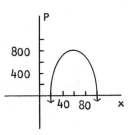

2.

4.

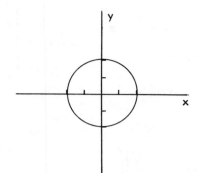

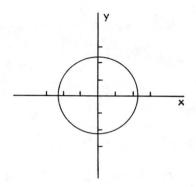

6.

8.

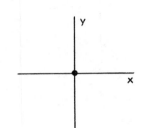

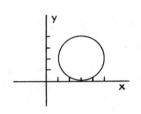

10.

12.

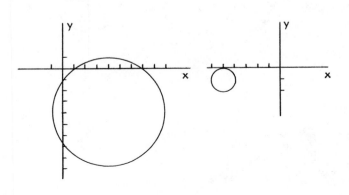

14. $x^2 + 4x + y^2 - 6x + 4 = 0$

$(x^2 + 4x \quad) + (y^2 - 6y \quad) = -4$

$(x^2 + 4x + 4) + (y^2 - 6y + 9) = -4 + 4 + 9$

$(x + 2)^2 + (y - 3)^2 = 9$

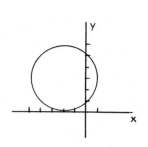

16. $x^2 - 10x + y^2 + 9 = 0$

$(x^2 - 10x \quad) + y^2 = -9$

$(x^2 - 10x + 25) + y^2 = -9 + 25$

$(x - 5)^2 + y^2 = 16$

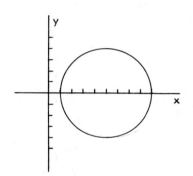

18. $(x - 3)^2 + (y - 6)^2 = 4^2$

$(x - 3)^2 + (y - 6)^2 = 16$

20. $(x - 5)^2 + [y - (-4)]^2 = \sqrt{6}\,^2$

$(x - 5)^2 + (y + 4)^2 = 6$

22. $(x - 0)^2 + [y - (-2)]^2 = 1^2$

$x^2 + (y + 2)^2 = 1$

24. Domain $= \{x \mid -3 \leq x \leq 3\}$

Range $= \{y \mid -3 \leq y \leq 0\}$

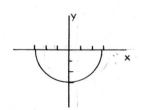

26. $x^2 + y^2 = (\frac{3}{2})^2$

$x^2 + y^2 = \frac{9}{4}$

Section 8.3

2. $9x^2 + 25y^2 = 225$

$$\frac{x^2}{25} + \frac{y^2}{9} = 1$$

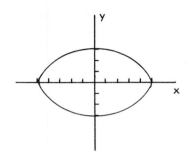

4. $25x^2 + 4y^2 = 100$

$$\frac{x^2}{4} + \frac{y^2}{25} = 1$$

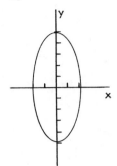

6. $x^2 + 16y^2 = 16$

$$\frac{x^2}{16} + \frac{y^2}{1} = 1$$

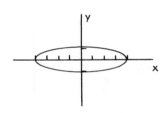

8. $7x^2 + 4y^2 = 28$

$$\frac{x^2}{4} + \frac{y^2}{7} = 1$$

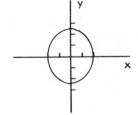

10. $x^2 - y^2 = 9$

$$\frac{x^2}{9} - \frac{y^2}{9} = 1$$

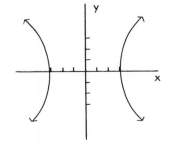

12. $9x^2 - 4y^2 = 36$

$$\frac{x^2}{4} - \frac{y^2}{9} = 1$$

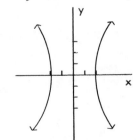

14. $x^2 - 9y^2 = 9$

$$\frac{x^2}{9} - \frac{y^2}{1} = 1$$

16. $4y^2 - 16x^2 = 16$

$$\frac{y^2}{4} - \frac{x^2}{1} = 1$$

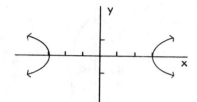

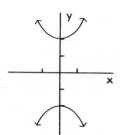

18. Domain $= \{x \mid x \neq 0\}$

    Range $= \{y \mid y \neq 0\}$

20. $y^2 - 4x^2 = 0$

    $(y - 2x)(y + 2x) = 0$

    $y - 2x = 0$ $y + 2x = 0$

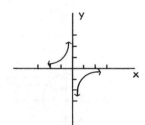

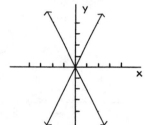

22. $(x + 3)^2 + (y - 3)^2 = 9$

24. $\frac{x^2}{9} + \frac{y^2}{25} = 1$

26. $\frac{y^2}{9} - x^2 = 1$

28. Demand never vanishes.

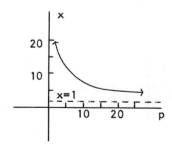

2.

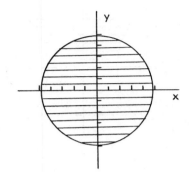

4.

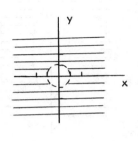

6.

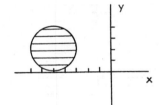

8.

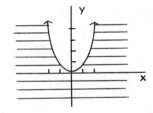

10.

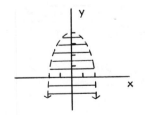

12.

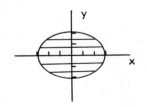

14.

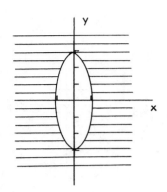

16.

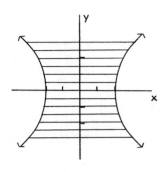

18.

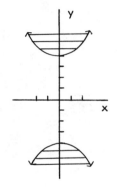

20.

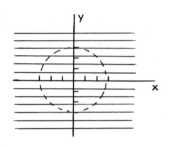

22. $x^2 - 6x + y^2 \geq 0$

$x^2 - 6x + 9 + y^2 \geq 9$

$(x - 3)^2 + y^2 \geq 9$

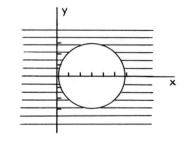

Section 8.5

2. $d = kw$

$5 = k6$

$k = \dfrac{5}{6}$

$d = \dfrac{5}{6}w$

$d = \dfrac{5}{6}(14) = 11\dfrac{2}{3}$ in.

4. $p = kd$

$65 = k15$

$k = \dfrac{13}{3}$

$p = \dfrac{13}{3}d$

$p = \dfrac{13}{3}(24) = 104$ lb/ft^2

6. $C = k\ell^2$

$60,000 = k(100)^2$

$k = 6$

$C = 6\ell^2$

$C = 6(200)^2 = \$240,000$

8. $R = \dfrac{k}{d^2}$

$0.5 = \dfrac{k}{(0.02)^2}$

$k = 0.0002$

$R = \dfrac{0.0002}{d^2}$

$R = \dfrac{0.0002}{(0.01)^2} = 2$ ohm

10. $w = \dfrac{kbd^2}{\ell}$

$1000 = \dfrac{k(1)(2)^2}{15}$

$k = 3750$

$w = \dfrac{3750bd^2}{\ell}$

$w = \dfrac{3750(1.5)(3)^2}{15}$

$= 3375$ lb

12. $A = kr^2$

Replace r by 2r and get an area of

$k(2r)^2 = 4kr^2$,

which is 4 times the original

area.

Section 8.6

2. $f^{-1} = \{(0, 1), (-2, 3), (4, 0)\}$

4. $f^{-1} = \{(0, 0), (1, 1), (2, 2)\} = f$

103

6. $f(x) = 4x + 1$

      $y = 4x + 1$

   f^{-1} is given by

      $x = 4y + 1$

   $x - 1 = 4y$

      $y = \dfrac{x - 1}{4}$

   $f^{-1}(x) = \dfrac{x - 1}{4}$.

8. $f(x) = x - 3$

      $y = x - 3$

   f^{-1} is given by

      $x = y - 3$

      $y = x + 3$

   $f^{-1}(x) = x + 3$.

10. $f(x) = x^2 - 2$

      $y = x^2 - 2$

   f^{-1} is given by

      $x = y^2 - 2$

      $y^2 = x + 2$

      $y = \overset{+}{-}\sqrt{x + 2}$

   $f^{-1}(x) = \overset{+}{-}\sqrt{x + 2}$.

12. $f(x) = x^3$

      $y = x^3$

   f^{-1} is given by

      $x = y^3$

      $y = \sqrt[3]{x}$

   $f^{-1}(x) = \sqrt[3]{x}$.

14. $f(x) = \sqrt{9 - x^2}$

      $y = \sqrt{9 - x^2}$,   where $-3 \leq x \leq 3$ and $y \geq 0$.

   f^{-1} is given by

      $x = \sqrt{9 - y^2}$,   where $-3 \leq y \leq 3$ and $x \geq 0$.

      $x^2 = 9 - y^2$

      $y^2 = 9 - x^2$

      $y = \overset{+}{-}\sqrt{9 - x^2}$,   where $-3 \leq y \leq 3$ and $x \geq 0$.

16. $f(x) = -1$

      $y = -1$

   f^{-1} is given by

      $x = -1$.

18. $f(x) = 3^x$

      $y = 3^x$

   f^{-1} is given by

      $x = 3^y$.

20. $g(x) = \frac{1}{3}x + 2$

$y = \frac{1}{3}x + 2$

g^{-1} is given by

$x = \frac{1}{3}y + 2$

$x - 2 = \frac{1}{3}y$

$y = 3x - 6$

$g^{-1}(x) = 3x - 6$.

Therefore,

$g(g^{-1}(2)) = g(3\cdot 2 - 6) = g(0) = \frac{1}{3}\cdot 0 + 2 = 2$

and

$g^{-1}(g(4)) = g^{-1}(\frac{1}{3}\cdot 4 + 2) = g^{-1}(\frac{10}{3}) = 3\cdot\frac{10}{3} - 6 = 4$.

In general,

$g(g^{-1}(x)) = g^{-1}(g(x)) = x$.

22.   24.

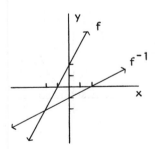

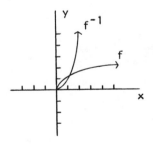

CHAPTER NINE
EXPONENTIAL AND LOGARITHMIC FUNCTIONS

Section 9.1

2.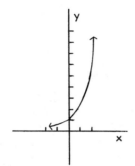

4.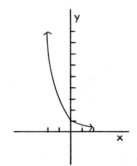

6. Same as Problem 4.

8.

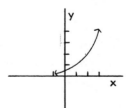

10.

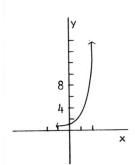

12.

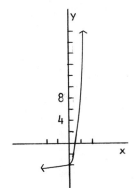

14.

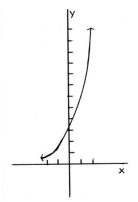

16.

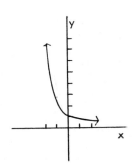

18.

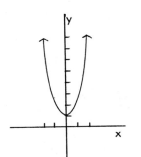

20.

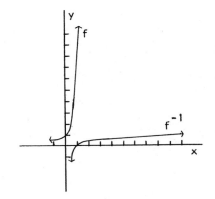

22. Some points of the graph lie below.

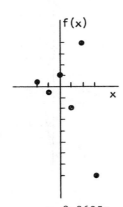

24. $A(t) = 20,000e^{0.0625t}$

$A(4) = 20,000e^{0.0625(4)}$

$= 20,000e^{0.25}$

$\approx 20,000 \ (1.2840)$

$\approx \$25,680$

26. $A(t) = A_0 2^{-0.05t}$

$A(t) = 600 \cdot 2^{-0.05t}$
$A(60) = 600 \cdot 2^{-0.05(60)}$

$= 600 \cdot 2^{-3}$

$= 75 \ g$

$A(10) = 600 \cdot 2^{-0.05(10)}$

$= 600 \cdot 2^{-0.5}$

$= 600 \cdot \dfrac{1}{\sqrt{2}}$

$= 300 \ \sqrt{2}$

$\approx 424.2 \ g$

28. $N(h) = 5 \cdot 3^{h}$

$N(8) = 5 \cdot 3^{8}$

$= 32,805 \ cells$

30. $P(h) = 14.7e^{-0.21h}$

$P(-5) = 14.7e^{-0.21(-5)}$

$= 14.7e^{1.05}$

$\approx 14.7(2.8577)$

$\approx 42 \ lb/in^{2}$

32. $P(n) = 5,000,000 \ (1.02)^n$

    $P(3) = 5,000,000 \ (1.061208)$

         $= 5,306,040$ persons

Section 9.2

2. $4^2 = 16$

4. $9^{\frac{1}{2}} = 3$

6. $e^0 = 1$

8. $10^1 = 10$

10. $10^{-3} = 0.001$

12. $(\frac{1}{3})^{-4} = 81$

14. $2^{-5} = \frac{1}{32}$

16. $\log_6 36 = 2$

18. $\log_{10} 0.1 = -1$

20. $\log_e 1 = 0$

22. $\log_{\frac{1}{2}} \frac{1}{8} = 3$

24. $\log_{27} 9 = \frac{2}{3}$

26. $\log_4 2 = -\frac{1}{2}$

28. $\log_{81} 9 = \frac{1}{2}$

30. $\log_b t = s$

32. 4

34. 7

36. $\frac{1}{2}$

38. $\frac{1}{2}$

40. 0

42. -2

44. 5

46. $y = \log_{10} 0.0001$

      $10^y = 0.0001$

      $y = -4$

48. $\log_4 x = \frac{5}{2}$

      $4^{\frac{5}{2}} = x$

      $x = 32$

50. $\log_a 0.1 = -1$

      $a^{-1} = 0.1$

      $a^{-1} = \frac{1}{10}$

      $a \ \ = 10$

52. $y = \log_2 \sqrt[3]{2}$

      $2^y = \sqrt[3]{2}$

      $2^y = 2^{1/3}$

      $y \ \ = \frac{1}{3}$

54. $y = \log_e e \sqrt[3]{e}$

$\quad e^y = e \sqrt[3]{e}$

$\quad e^y = e^{4/3}$

$\quad y = \dfrac{4}{3}$

56. $y = \log_4 x$

$\quad x = 4^y$

58. $y = \log_{\frac{1}{4}} x$

$\quad x = (\dfrac{1}{4})^y$

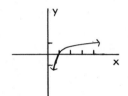

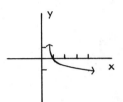

60. Undefined, since there is no real solution to the exponential equation $10^y = -1$.

62. Undefined, since there is no real solution to the exponential equation $0^y = 5$.

Section 9.3

2. $\log_a xy^3 = \log_a x + \log_a y^3 = \log_a x + 3 \log_a y$

4. $\log_a \dfrac{x}{y^2} = \log_a x - \log_a y^2 = \log_a x - 2 \log_a y$

6. $\log_a \dfrac{x\sqrt{y}}{z^3} = \log_a x + \log_a y^{\frac{1}{2}} - \log_a z^3$

$\qquad\qquad = \log_a x + \dfrac{1}{2} \log_a y - 3 \log_a z$

8. $\log_a \sqrt{x\sqrt{yz}} = \log_a [x(yz)^{\frac{1}{2}}]^{\frac{1}{2}}$

$$= \frac{1}{2} \log_a xy^{\frac{1}{2}}z^{\frac{1}{2}}$$

$$= \frac{1}{2} [\log_a x + \log_a y^{\frac{1}{2}} + \log_a z^{\frac{1}{2}}]$$

$$= \frac{1}{2} \log_a x + \frac{1}{4} \log_a y + \frac{1}{4} \log_a z$$

10. $3 \log_a x + \log_a y - \log_a z = (\log_a x^3 + \log_a y) - \log_a z$

$$= \log_a \frac{x^3 y}{z}$$

12. $\log_a x - \frac{1}{2} \log_a y - 3 \log_a z = \log_a x - (\log_a y^{\frac{1}{2}} + \log_a z^3)$

$$= \log_a \frac{x}{\sqrt{y} \; z^3}$$

14. $\log_{10}(x + 1) + \log_{10}(x + 2) = \log_{10}(x + 1)(x + 2) = \log_{10}(x^2+3x+2)$

16. $\log_e xy^2 + 2 \log_e \frac{x}{y} = \log_e xy^2 + \log_e (\frac{x}{y})^2$

$$= \log_e xy^2 \cdot \frac{x^2}{y^2}$$

$$= \log_e x^3$$

18. $\log_{10} 6 = \log_{10} 2 \cdot 3 = \log_{10} 2 + \log_{10} 3 = 0.3010 + 0.4771 = 0.7781$

20. $\log_{10} 30 = \log_{10} 2 \cdot 3 \cdot 5$

$$= \log_{10} 2 + \log_{10} 3 + \log_{10} 5$$

$$= 0.3010 + 0.4771 + 0.6990$$

$$= 1.4771$$

22. $\log_{10} 1.5 = \log_{10} \frac{3}{2} = \log_{10} 3 - \log_{10} 2 = 0.4771-0.3010 = 0.1761$

24. $\log_{10} 0.4 = \log_{10} \frac{2}{5} = \log_{10} 2 - \log_{10} 5 = 0.3010 - 0.6990 = -0.3980$

26. $\log_{10} \sqrt{2} = \log_{10} 2^{\frac{1}{2}} = \frac{1}{2} \log_{10} 2 = \frac{1}{2} (0.3010) = 0.1505$

28. $\log_{10} \sqrt[3]{5} = \log_{10} 5^{\frac{1}{3}} = \frac{1}{3} \log_{10} 5 = \frac{1}{3} (0.6990) = 0.2330$

30. $\log_{10} 900 = \log_{10} (9 \times 10^2)$

$$= \log_{10} 3^2 + \log_{10} 10^2$$

$$= 2 \log_{10} 3 + 2$$

$$= 2 (0.4771) + 2$$

$$= 2.9542$$

32. $\log_{10} 0.0006 = \log_{10} (6 \times 10^{-4})$

$$= \log_{10} 2 \cdot 3 + \log_{10} 10^{-4}$$

$$= \log_{10} 2 + \log_{10} 3 + (-4)$$

$$= 0.3010 + 0.4771 - 4$$

$$= -3.2219$$

34. $\log_2 \frac{16}{2} = \log_2 8 = 3$ but $\dfrac{\log_2 16}{\log_2 2} = \dfrac{4}{1} = 4.$

36. Let $r \quad = \log_a N$

Then $a^r \quad = N$

$\qquad (a^r)^c = N^c$

$\qquad a^{cr} \quad = N^c$

Therefore,

$$\log_a N^c = \log_a a^{cr} = cr = c \log_a N$$

Section 9.4

2. Since $10^2 < 343 < 10^3$, then $2 < \log 343 < 3.$

112

4. Since $10^1 < 45 < 10^2$, then $1 < \log 45 < 2$.

6. Since $10^0 < 7.2 < 10^1$, then $0 < \log 7.2 < 1$.

8. Since $10^{-2} < 0.08 < 10^{-1}$, then $-2 < \log 0.08 < -1$.

10. $\log 5.43 = 0.7348$

12. $\log 2.07 = 0.3160$

14. $\log 3.9 = 0.5911$

16. $\log 699 = \log (6.99 \times 10^2)$

$$= \log 6.99 + \log 10^2$$
$$= 0.8445 + 2$$
$$= 2.8445$$

18. $\log 84 = \log (8.4 \times 10)$

$$= \log 8.4 + \log 10$$
$$= 0.9243 + 1$$
$$= 1.9243$$

20. $\log 43,700 = \log (4.37 \times 10^4)$

$$= \log 4.37 + \log 10^4$$
$$= 0.6405 + 4$$
$$= 4.6405$$

22. $\log 24.3 = \log (2.43 \times 10)$

$$= \log 2.43 + \log 10$$
$$= 0.3856 + 1$$
$$= 1.3856$$

24. $\log 0.0519 = \log (5.19 \times 10^{-2})$

$$= \log 5.19 + \log 10^{-2}$$
$$= 0.7152 + (-2)$$
$$= 0.7152 - 2$$

26. $\log 0.00676 = \log (6.76 \times 10^{-3})$

$\quad\quad\quad\quad = \log 6.76 + \log 10^{-3}$

$\quad\quad\quad\quad = 0.8299 - 3$

28. $\log 0.505 = \log (5.05 \times 10^{-1})$

$\quad\quad\quad\quad = \log 5.05 + \log 10^{-1}$

$\quad\quad\quad\quad = 0.7033 - 1$

30. $\log 0.0009 = \log (9 \times 10^{-4})$

$\quad\quad\quad\quad = \log 9 + \log 10^{-4}$

$\quad\quad\quad\quad = 0.9542 - 4$

32. $x = 1.65 \times 10^{0} = 1.65$　　　34. $x = 3.08 \times 10^{3} = 3080$

36. $x = 3.55 \times 10^{2} = 355$　　　38. $x = 8.13 \times 10^{-2} = 0.0813$

40. $x = 3.25 \times 10^{-3} = 0.00325$

42. $\log x = -5.2890$　　　　　　44. $\log x = -2.3288$

$\quad\quad\quad = -5.2890 + 6 - 6$　　　　　　$= -2.3288 + 3 - 3$

$\quad\quad\quad = 0.7110 - 6$　　　　　　　$= 0.6712 - 3$

$\quad\quad x = 5.14 \times 10^{-6}$　　　　　　$x = 4.69 \times 10^{-3}$

$\quad\quad\quad = 0.00000514$　　　　　　　$= 0.00469$

Section 9.5

2.

$$10 \left\{ \begin{array}{l} \log 5.740 = 0.7589 \\ 4 \left\{ \begin{array}{l} \log 5.734 = ? \\ \log 5.730 = 0.7582 \end{array} \right\} d \end{array} \right\} 0.0007$$

$$\frac{d}{0.0007} = \frac{4}{10}$$

$$d = 0.0003$$

$$\log 5.734 = 0.7582 + 0.0003 = 0.7585$$

4. $\log 86.64 = \log (8.664 \times 10^1) = \log 8.664 + 1$

$$10\left\{\,4\left\{\begin{array}{l}\log 8.670 = 0.9380 \\ \log 8.664 = \ ? \\ \log 8.660 = 0.9375\end{array}\right\}d\ \right\}0.0005$$

$$\frac{d}{0.0005} = \frac{4}{10}$$

$$d = 0.0002$$

$$\log 8.664 = 0.9375 + 0.0002 = 0.9377$$

$$\log 86.64 = 0.9377 + 1 = 1.9377$$

6. $\log 564.7 = \log 5.647 + 2$

$$10\left\{\,7\left\{\begin{array}{l}\log 5.650 = 0.7520 \\ \log 5.647 = \ ? \\ \log 5.640 = 0.7513\end{array}\right\}d\ \right\}0.0007$$

$$\frac{d}{0.0007} = \frac{7}{10}$$

$$d = 0.0005$$

$$\log 5.647 = 0.7518$$

$$\log 564.7 = 2.7518$$

8. $\log 137{,}800 = \log 1.378 + 5$

$$10\left\{\,8\left\{\begin{array}{l}\log 1.380 = 0.1399 \\ \log 1.378 = \ ? \\ \log 1.370 = 0.1367\end{array}\right\}d\ \right\}0.0032$$

$$\frac{d}{0.0032} = \frac{8}{10}$$

$$d = 0.0026$$

$$\log 1.378 = 0.1393$$

$$\log 137{,}800 = 5.1393$$

10. $\log 0.8634 = \log 8.634 + (-1)$

$$10 \left\{ \begin{array}{l} \log 8.640 = 0.9365 \\ 4 \left\{ \begin{array}{l} \log 8.634 = ? \\ \log 8.630 = 0.9360 \end{array} \right\} d \end{array} \right\} 0.0005$$

$$\frac{d}{0.0005} = \frac{4}{10}$$

$$d = 0.0002$$

$$\log 8.634 = 0.9362$$

$$\log 0.8634 = 0.9362 - 1$$

12. $\log 0.003766 = \log 3.766 + (-3)$

$$10 \left\{ \begin{array}{l} \log 3.770 = 0.5763 \\ 6 \left\{ \begin{array}{l} \log 3.766 = ? \\ \log 3.760 = 0.5752 \end{array} \right\} d \end{array} \right\} 0.0011$$

$$\frac{d}{0.0011} = \frac{6}{10}$$

$$d = 0.0007$$

$$\log 3.766 = 0.5759$$

$$\log 0.003766 = 0.5759 - 3$$

14.

$$0.01 \left\{ \begin{array}{l} \log 3.55 = 0.5502 \\ d \left\{ \begin{array}{l} \log x \quad = 0.5495 \\ \log 3.54 = 0.5490 \end{array} \right\} 5 \end{array} \right\} 12$$

$$\frac{d}{0.01} = \frac{5}{12}$$

$$d = 0.004$$

$$x = 3.54 + 0.004 = 3.544$$

16.

$$0.1 \left\{ \begin{array}{l} \log 28.3 = 1.4518 \\ d \left\{ \begin{array}{l} \log x \quad = 1.4510 \\ \log 28.2 = 1.4502 \end{array} \right\} 8 \end{array} \right\} 16$$

$$\frac{d}{0.1} = \frac{8}{16}$$

$$d = 0.05$$

$$x = 28.2 + 0.05 = 28.25$$

18.

$$10 \left\{ \begin{array}{l} \log 3730 = 3.5717 \\ d \left\{ \begin{array}{l} \text{1ox } x \quad = 3.5711 \\ \log 3720 = 3.5705 \end{array} \right\} 6 \end{array} \right\} 12$$

$$\frac{d}{10} = \frac{6}{12}$$

$$d = 5$$

$$x = 3725$$

20.

$$0.0001 \left\{ \begin{array}{l} \log 0.0184 = 0.2648 - 2 \\ d \left\{ \begin{array}{l} \log x \quad = 0.2631 - 2 \\ \log 0.0183 = 0.2625 - 2 \end{array} \right\} 6 \end{array} \right\} 23$$

$$\frac{d}{0.0001} = \frac{6}{23}$$

$$d = 0.00003$$

$$x = 0.01833$$

22.

$$0.001 \left\{ \begin{array}{l} \log 0.164 = 0.2148 - 1 \\ d \left\{ \begin{array}{l} \log x \quad = 0.2128 - 1 \\ \log 0.163 = 0.2122 - 1 \end{array} \right\} 6 \end{array} \right\} 26$$

$$\frac{d}{0.001} = \frac{6}{26}$$

$$d = 0.0002$$

$$x = 0.1632$$

24. $\log x = -1.1677 = -1.1677 + 2 - 2 = 0.8323 - 2$

$$0.0001 \left\{ \begin{array}{l} \begin{array}{l} \log 0.0680 = 0.8325 - 2 \\ d \left\{ \begin{array}{l} \log x \quad\;\; = 0.8323 - 2 \\ \log 0.0679 = 0.8319 - 2 \end{array} \right\} 4 \end{array} \end{array} \right\} 6$$

$$\frac{d}{0.0001} = \frac{4}{6}$$

$$d = 0.00007$$

$$x = 0.06797$$

Section 9.6

2. $\quad x = (49.3)(0.515)$

$\log \; x = \log \; (49.3)(0.515)$

$\quad = \log \; 49.3 + \log \; 0.515$

$\quad = 1.6928 + (0.7118 - 1)$

$\quad = 1.4046$

$x = 25.4$

4. $\quad x = (0.00527)(3.05)$

$\log \; x = \log \; 0.00527 + \log \; 3.05$

$\quad = (0.7218 - 3) + .4843$

$\quad = 0.2061 - 2$

$x = 0.0161$

6. $\quad x = \dfrac{95.3}{6.75}$

$\log \; x = \log \; \dfrac{95.3}{6.75}$

$\quad = \log \; 95.3 - \log \; 6.75$

$\quad = 1.9791 - 0.8293$

$\quad = 1.1498$

$x = 14.1$

8. $x = \dfrac{0.000645}{0.00914}$

log x = log 0.000645 - log 0.00914

$= (0.8096 - 4) - (0.9609 - 3)$

$= 0.8487 - 2$

$x = 0.0706$

10. $x = \dfrac{(83,200)(0.00414)}{73.6}$

log x = log 83,200 + log 0.00414 - log 73.6

$= 4.9201 + (0.6170 - 3) - 1.8669$

$= 0.6702$

$x = 4.68$

12. $x = (1.75)^6$

log x = log $(1.75)^6$

$= 6 \text{ log } 1.75$

$= 6(0.2430)$

$= 1.4580$

$x = 28.7$

14. $x = \sqrt[5]{619} = (619)^{\frac{1}{5}}$

log x = $\dfrac{1}{5}$ log 619

$= \dfrac{1}{5} (2.7917)$

$= 0.5583$

$x = 3.62$

16. $x = (15.2)^{1.3}$

log x = 1.3 log 15.2

$= 1.3 (1.1818)$

$= 1.5363$

$x = 34.4$

18. $\quad x = \dfrac{(3.76)(0.873)^3}{2.03}$

$\log\ x = \log 3.76 + 3 \log 0.873 - \log 2.03$

$\qquad = 0.5752 + 3\,(0.9410 - 1) - 0.3075$

$\qquad = 0.0907$

$\quad x = 1.23$

20. $\quad x = \dfrac{(2.07)^3 (8.15)}{\sqrt{941}}$

$\log\ x = 3 \log 2.07 + \log 8.15 - \dfrac{1}{2} \log 941$

$\qquad = 3\,(0.3160) + (0.9112) - \dfrac{1}{2}\,(2.9736)$

$\qquad = 0.3724$

$\quad x = 2.36$

22. $\quad x = \sqrt[3]{\dfrac{(4.09)(0.677)}{1.26}}$

$\log\ x = \dfrac{1}{3} \left[\log 4.09 + \log 0.677 - \log 1.26 \right]$

$\qquad = \dfrac{1}{3} \left[0.6117 + (0.8306 - 1) - 0.1004 \right]$

$\qquad = 0.1140$

$\quad x = 1.30$

24. $\quad x = \dfrac{87,420}{68.21}$

$\log\ x = \log 87,420 - \log 68.21$

$\qquad = 4.9416 - 1.8338$

$\qquad = 3.1078$

$\quad x = 1282$

26. $x = \sqrt{5146}$

$\log x = \dfrac{1}{2} \log 5146$

$ = \dfrac{1}{2} (3.7115)$

$ = 1.8557$

$x = 71.73$

28. $T = 2\pi\sqrt{\dfrac{\ell}{g}}$

$T = 2\,(3.14)\sqrt{\dfrac{3}{32.2}}$

$T = 6.28\sqrt{\dfrac{3}{32.2}}$

$\log T = \log 6.28 + \dfrac{1}{2}\left[\log 3 - \log 32.2\right]$

$ = 0.7980 + \dfrac{1}{2}\left[0.4771 - 1.5079\right]$

$ = 0.2826$

$T = 1.92 \text{ sec}$

30. $V = \dfrac{4}{3}\pi r^3$

$3V = 4\pi r^3$

$r^3 = \dfrac{3V}{4\pi}$

$r = \sqrt[3]{\dfrac{3V}{4\pi}}$

$r = \sqrt[3]{\dfrac{3(12)}{4(3.14)}}$

$r = \sqrt[3]{\dfrac{9}{3.14}}$

$\log r = \dfrac{1}{3}(\log 9 - \log 3.14)$

$ = \dfrac{1}{3}(0.9542 - 0.4969)$

$ = 0.1524$

$r = 1.42 \text{ cm}$

2. $3^x = 5$

$\log 3^x = \log 5$

$x \log 3 = \log 5$

$x = \dfrac{\log 5}{\log 3}$

$x = 1.465$

4. $6^{3x} = 11$

$\log 6^{3x} = \log 11$

$3x \log 6 = \log 11$

$3x = \dfrac{\log 11}{\log 6}$

$x = \dfrac{\log 11}{3 \log 6} = 0.4461$

6. $2^{x-1} = 7$

$\log 2^{x-1} = \log 7$

$(x - 1) \log 2 = \log 7$

$x - 1 = \dfrac{\log 7}{\log 2}$

$x = \dfrac{\log 7}{\log 2} + 1$

$x = 3.807$

8. $5^{2x-1} = 15$

$\log 5^{2x-1} = \log 15$

$(2x - 1)\log 5 = \log 15$

$2x - 1 = \dfrac{\log 15}{\log 5}$

$2x = \dfrac{\log 15}{\log 5} + 1$

$x = \dfrac{1}{2} \left(\dfrac{\log 15}{\log 5} + 1\right)$

$x = 1.341$

10. $3^{x^2} = 100$

$\log 3^{x^2} = \log 100$

$x^2 \log 3 = 2$

$x^2 = \dfrac{2}{\log 3}$

$x^2 = 4.192$

$x = \pm 2.047$

12. $\log (x + 1) = 3$

$x + 1 = 10^3$

$x + 1 = 1000$

$x = 999$

14. $\log 4 + \log x = 2$

$\qquad \log 4x = 2$

$\qquad 4x = 10^2$

$\qquad 4x = 100$

$\qquad 4x = 25$

16. $\log x + \log (x + 21) = 2$

$\qquad \log x(x + 21) = 2$

$\qquad x(x + 21) = 10^2$

$\qquad x^2 + 21x - 100 = 0$

$\qquad (x + 25)(x - 4) = 0$

$\qquad x = -25 \qquad x = 4$

But $x = -25$ does not check, since $\log (-25)$ and $\log (-4)$ are undefined. Therefore, the solution set is $\{4\}$.

18. $\log (x + 1) - \log (x - 1) = 1$

$\qquad \log \dfrac{x + 1}{x - 1} = 1$

$\qquad \dfrac{x + 1}{x - 1} = 10^1$

$\qquad x + 1 = 10x - 10$

$\qquad 11 = 9x$

$\qquad \dfrac{11}{9} = x$

20. $\qquad \log (2x - 1) = \log 9$

$\log (2x - 1) - \log 9 = 0$

$\qquad \log \dfrac{2x - 1}{9} = 0$

$\qquad \dfrac{2x - 1}{9} = 10^0$

$\qquad 2x - 1 = 9$

$\qquad x = 5$

22. $N(h) = 10 \cdot 3^h$

$\qquad 60 = 10 \cdot 3^h$

$\qquad 6 = 3^h$

$\qquad \log 6 = \log 3^h$

$\qquad \log 6 = h \log 3$

$\qquad h = \dfrac{\log 6}{\log 3}$

$\qquad h = 1.631 \text{ hr}$

24. $pH = -\log(5.8 \times 10^{-10})$

$\qquad = -(\log 5.8 + \log 10^{-10})$

$\qquad = -(0.7634 - 10)$

$\qquad = 9.2$

26. $A(t) = A_0 2^{-\frac{t}{5700}}$

$\qquad .2A_0 = A_0 2^{-\frac{t}{5700}}$

$\qquad .2 = 2^{-\frac{t}{5700}}$

$\qquad \log .2 = \log 2^{-\frac{t}{5700}}$

$\qquad \log .2 = -\dfrac{t}{5700} \log 2$

$\qquad \dfrac{\log .2}{\log 2} = -\dfrac{t}{5700}$

$\qquad t = -5700 \left(\dfrac{\log .2}{\log 2}\right)$

$\qquad t = 13,200 \text{ yr}$

28. $P(n) = 2(1.01)^n$

$\qquad 6 = 2(1.01)^n$

$\qquad 3 = (1.01)^n$

$\qquad \log 3 = \log (1.01)^n$

$\qquad \log 3 = n \log 1.01$

$\qquad n = \dfrac{\log 3}{\log 1.01}$

$\qquad n = 110 \text{ yr}$

30. $\qquad y = A \cdot 10^{-2x}$

$\qquad \dfrac{y}{A} = 10^{-2x}$

$\qquad \log \dfrac{y}{A} = \log 10^{-2x}$

$\qquad \log \dfrac{y}{A} = -2x \log 10$

$\qquad \log \dfrac{y}{A} = -2x \cdot 1$

$\qquad x = -\dfrac{1}{2} \log \dfrac{y}{A}$

Section 9.8

2. $\log_2 5 = \dfrac{\log 5}{\log 2} = 2.32$ 4. $\log_6 12 = \dfrac{\log 12}{\log 6} = 1.39$

6. $\log_{50} 100 = \dfrac{\log 100}{\log 50} = 1.18$ 8. $\ln 3 = \dfrac{\log 3}{\log e} = 1.10$

10. $\ln 75 = \dfrac{\log 75}{\log e} = 4.32$ 12. $\log_{\frac{1}{2}} 10 = \dfrac{\log 10}{\log \frac{1}{2}} = -3.32$

14. $\log_3 11 = \dfrac{\log_8 11}{\log_8 3}$ 16. $\log_6 2 = \dfrac{1}{\log_2 6}$

18. $\log_{26} 10 = \dfrac{1}{\log 26} = \dfrac{1}{1.4150} = 0.707$

CHAPTER TEN
SYSTEMS OF EQUATIONS AND INEQUALITIES

Section 10.1

2. Independent

4. Inconsistent

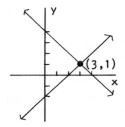

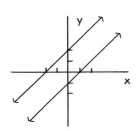

6. Dependent

8. Independent

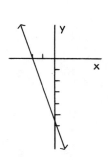

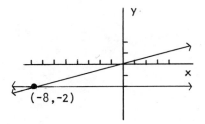

$2y = 6$
$2y = 10$

12.
$3x - y = 2$
$5x + y = 6$

$x = 6 + 2y$

$y = 6 - 5x$

$?y = 10$

$3x - (6 - 5x) = 2$

$y = 4$

$3x - 6 + 5x = 2$

$? = 1$

$8x = 8$

$= 6 + 2(1)$

$x = 1$

$= 8$

$y = 6 - 5(1)$

$= 1$

14.
$2x + 3y = 6$
$4x - 3y = -4$

16.
$2x + 3y = 3$
$4x + 5y = 5$

$2x = 6 - 3y$

$2x = 3 - 3y$

$x = \dfrac{6 - 3y}{2}$

$x = \dfrac{3 - 3y}{2}$

$4\left(\dfrac{6 - 3y}{2}\right) - 3y = -4$

$4\left(\dfrac{3 - 3y}{2}\right) + 5y = 5$

$12 - 6y - 3y = -4$

$6 - 6y + 5y = 5$

$-9y = -16$

$-y = -1$

$y = \dfrac{16}{9}$

$y = 1$

$x = \dfrac{6 - 3\left(\frac{16}{9}\right)}{2}$

$x = \dfrac{3 - 3(1)}{2}$

$= \dfrac{1}{3}$

$= 0$

18. $x + y = 8$
$\underline{x - y = 2}$
$2x \quad\;\; = 10$

$x = 5$

$5 + y = 8$

$y = 3$

127

20. $2x + 3y = 2$
    $4x + 9y = 9$

    $-4x - 6y = -4$
    $\underline{4x + 9y = 9}$
           $3y = 5$

$$y = \frac{5}{3}$$

$$2x + 3\left(\frac{5}{3}\right) = 2$$

$$2x = -3$$

$$x = -\frac{3}{2}$$

22. $3x - 2y = 0$
    $7x - 3y = 5$

    $9x - 6y = 0$
    $\underline{-14x + 6y = -10}$
    $-5x \quad\quad = -10$

$$x = 2$$

$$3(2) - 2y = 0$$

$$-2y = -6$$

$$y = 3$$

24. $-\frac{1}{4}x + y = 2$

    $\frac{3}{4}x + \frac{1}{2}y = 1$

    $-3x + 12y = 24$
    $\underline{3x + 2y = 4}$
         $14y = 28$

$$y = 2$$

$$-\frac{1}{4}x + 2 = 2$$

$$-\frac{1}{4}x = 0$$

$$x = 0$$

26. $x + y = 1$
    $ax - by = 1$

    $bx + by = b$
    $\underline{ax - by = 1}$
    $(b + a)x \quad\quad = b + 1$

$$x = \frac{b + 1}{b + a}$$

$$\frac{b + 1}{b + a} + y = 1$$

$$y = 1 - \frac{b + 1}{b + a}$$

$$y = \frac{a - 1}{b + a}$$

28. $w = \text{width}$

    $\ell = \text{length}$

    $w = \frac{1}{3}\ell$

    $w + \ell = 16$

$$\frac{1}{3}\ell + \ell = 16$$

$$\ell + 3\ell = 48$$

$$4\ell = 48$$

$$\ell = 12 \text{ cm}$$

$$w = \frac{1}{3}(12) = 4 \text{ cm}$$

30. d = no. of dimes

q = no. of quarters

$$d = 1 + 2q$$
$$10d + 25q = 235$$

$$10(1 + 2q) + 25q = 235$$

$$10 + 20q + 25q = 235$$

$$45q = 225$$

$$q = 5 \text{ quarters}$$

$$d = 1 + 2(5)$$

$$= 11 \text{ dimes}$$

32. $y = mx + b$

(-3, 4): $4 = m(-3) + b$ $\Rightarrow$ $-3m + b = 4$
(3,-8): $-8 = m(3) + b$ $\Rightarrow$ $\underline{3m + b = -8}$
$$2b = -4$$

$$b = -2$$

$$3m - 2 = -8$$

$$3m = -6$$

$$m = -2$$

34. $x = 60 - 2p$
      $x = 3p - 20$

$$-x = -60 + 2p$$
$$\underline{x = -20 + 3p}$$
$$0 = -80 + 5p$$

$$-5p = -80$$

$$p = 16$$

2. $x + y + z = 4$
    $\underline{x - 2y - z = 1}$
    $2x - y \quad\;\; = 5$

    $2x + 2y + 2z = 8$
    $\underline{2x - y - 2z = -1}$
    $4x + y \qquad\; = 7$

    $2x - y \quad\;\; = 5$
    $\underline{4x + y \quad\;\; = 7}$
    $6x \qquad\;\; = 12$

    $x = 2$

    $4(2) + y = 7$

    $y = -1$

    $2 - 1 + z = 4$

    $z = 3$

4. $x + y + z = 6$
    $\underline{2x + 3y - z = 7}$
    $3x + 4y \qquad = 13$

    $x + y + z = 6$
    $\underline{3x - y - z = 6}$
    $4x \qquad\quad = 12$

    $x = 3$

    $3(3) + 4y = 13$

    $4y = 4$

    $y = 1$

    $3 + 1 + z = 6$

    $z = 2$

6. $\begin{array}{rcrcrcr} 2x & - & 4y & + & 2z & = & -2 \\ -2x & - & y & + & 3z & = & -3 \\ \hline & & -5y & + & 5z & = & -5 \end{array}$ $\begin{array}{rcrcrcr} -3x & + & 6y & - & 3z & = & 3 \\ 3x & + & 3y & - & 2z & = & 10 \\ \hline & & 9y & - & 5z & = & 13 \end{array}$

$$\begin{array}{rcrcr} -5y & + & 5z & = & -5 \\ 9y & - & 5z & = & 13 \\ \hline 4y & & & = & 8 \end{array}$$

$$y = 2$$

$$-5(2) + 5z = -5$$

$$5z = 5$$

$$z = 1$$

$$2x-4(2)+2(1) = -2$$

$$2x = 4$$

$$x = 2$$

8. $\begin{array}{rcrcrcr} 2x & + & 2y & - & 7z & = & -3 \\ 8x & - & 2y & + & 6z & = & -8 \\ \hline 10x & & & - & z & = & -11 \end{array}$ $\begin{array}{rcrcrcr} 8x & - & 2y & + & 6z & = & -8 \\ 5x & + & 2y & - & 2z & = & -1 \\ \hline 13x & & & + & 4z & = & -9 \end{array}$

$$\begin{array}{rcrcr} 40x & - & 4z & = & -44 \\ 13x & + & 4z & = & -9 \\ \hline 53x & & & = & -53 \end{array}$$

$$x = \mathbf{-1}$$

$$13(-1) + 4z = -9$$

$$4z = 4$$

$$z = 1$$

$$2(-1) + 2y - 7(1) = -3$$

$$2y = 6$$

$$y = 3$$

10.
$$\begin{array}{rcr} y - z &=& 1 \\ -7y + z &=& -1 \\ \hline -6y &=& 0 \end{array} \qquad \begin{array}{rcr} 12x - 4y + 4z &=& 8 \\ -12x - 3y - 3z &=& -9 \\ \hline -7y + z &=& -1 \end{array}$$

$$y = 0$$

$$0 - z = 1$$

$$z = -1$$

$$3x - 0 - 1 = 2$$

$$3x = 3$$

$$x = 1$$

12.
$$\begin{array}{rcr} 4x - 2z &=& 9 \\ -2x + 2z &=& -6 \\ \hline 2x &=& 3 \end{array} \qquad \begin{array}{rcr} 2x - y &=& 5 \\ y - 2z &=& 1 \\ \hline 2x - 2z &=& 6 \end{array}$$

$$x = \frac{3}{2}$$

$$2\left(\frac{3}{2}\right) - 2z = 6$$

$$-2z = 3$$

$$z = -\frac{3}{2}$$

$$y - 2\left(-\frac{3}{2}\right) = 1$$

$$y = -2$$

14.
$$\begin{array}{rcr} x - y - z &=& 3 \\ x + y - z &=& 1 \\ \hline 2x - 2z &=& 4 \end{array} \qquad \begin{array}{rcr} 2x - 2y - 2z &=& 6 \\ x + 2y - z &=& 3 \\ \hline 3x - 3z &=& 9 \end{array}$$

$$\begin{array}{rcr} x - z &=& 4 \\ -x + z &=& -3 \\ \hline 0 &=& 1 \end{array}$$

The system is inconsistent. The solution set is ∅.

132

16. $\begin{aligned} x + y - z &= -1 \\ x - 2y + z &= 4 \\ \hline 2x - y &= 3 \end{aligned}$ $\begin{aligned} -2x - 2y + 2z &= 2 \\ 4x + y - 2z &= 1 \\ \hline 2x - y &= 3 \end{aligned}$

The system is dependent. Let x be any real number c. Then

$$2c - y = 3$$

$$-y = 3 - 2c$$

$$y = 2c - 3$$

$$c+(2c-3) - z = -1$$

$$3c - 3 - z = -1$$

$$-z = 2 - 3c$$

$$z = 3c - 2$$

There are an infinite number of ordered-triple solutions, and they are all of the form (c, 2c - 3, 3c - 2).

18. x = shortest side
    y = medium side
    z = longest side

$$\begin{aligned} x + y + z &= 14 \\ z &= x + y \\ z &= 2x + 1 \end{aligned}$$

$$\begin{aligned} x + y + z &= 14 \\ -x - y + z &= 0 \\ \hline 2z &= 14 \end{aligned}$$

$$z = 7 \text{ cm}$$

$$7 = 2x + 1$$

$$-2x = -6$$

$$x = 3 \text{ cm}$$

$$3 + y + 7 = 14$$

$$y = 4 \text{ cm}$$

20.
$$y = ax^2 + bx + c$$

$(2, 5)$: $\quad 5 = a(2)^2 + b(2) + c \quad \Rightarrow \quad 4a + 2b + c = 5$

$(0, -3)$: $\quad -3 = a(0)^2 + b(0) + c \quad \Rightarrow \qquad\qquad c = -3$

$(-1, -4)$: $\quad -4 = a(-1)^2 + b(-1) + c \Rightarrow a - b + c \quad = -4$

$\qquad 4a + 2b - 3 = 5 \qquad\qquad\qquad\qquad a - b - 3 \quad = -4$

$$4a + 2b = 8 \qquad\qquad\qquad a - b \quad = -1$$
$$\underline{2a - 2b = -2}$$
$$6a \qquad = 6$$

$$a = 1$$

$$1 - b = -1$$

$$-b = -2$$

$$b = 2$$

An equation for the parabola is $y = x^2 + 2x - 3$.

22. x = hundreds digit

   y = tens digit

   z = units digit

$x + y + z = 16 \qquad\qquad (100x + 10y + z) - (100z + 10y + x) = 99$
$\qquad\quad y = z + 3$
$\qquad x - z = 1 \qquad\qquad\qquad\qquad\qquad 99x - 99z = 99$

$\qquad\qquad\qquad\qquad\qquad\qquad\qquad\qquad\qquad\qquad x - z = 1$

$x + (z + 3) + z = 16$

$$x + 2z = 13$$
$$\underline{-x + z = -1}$$
$$3z = 12$$

$$z = 4$$

$$y = 4 + 3$$

$$= 7$$

$$x - 4 = 1$$

$$x = 5$$

The original number is 574.

Section 10.3

2. $\begin{vmatrix} 3 & 2 \\ 4 & 6 \end{vmatrix} = 18 - 8 = 10$

4. $\begin{vmatrix} 5 & 2 \\ 7 & 1 \end{vmatrix} = 5 - 14 = -9$

6. $\begin{vmatrix} -3 & 2 \\ 5 & -1 \end{vmatrix} = 3 - 10 = -7$

8. $\begin{vmatrix} 2 & -4 \\ -3 & 1 \end{vmatrix} = 2 - 12 = -10$

10. $\begin{vmatrix} 0 & 1 \\ 1 & 0 \end{vmatrix} = 0 - 1 = -1$

12. $\begin{vmatrix} 0 & 0 \\ -8 & 2 \end{vmatrix} = 0 - 0 = 0$

14. $\begin{vmatrix} -1 & 4 & -2 \\ 2 & 2 & 3 \\ 3 & -1 & 1 \end{vmatrix} = -2 \begin{vmatrix} 4 & -2 \\ -1 & 1 \end{vmatrix} + 2 \begin{vmatrix} -1 & -2 \\ 3 & 1 \end{vmatrix} - 3 \begin{vmatrix} -1 & 4 \\ 3 & -1 \end{vmatrix}$

$= -2(4 - 2) + 2[-1 - (-6)] - 3(1 - 12)$

$= -4 + 10 + 33$

$= 39$

16. $\begin{vmatrix} 1 & -2 & -1 \\ 3 & -1 & -3 \\ 6 & 2 & 5 \end{vmatrix} = -3 \begin{vmatrix} -2 & -1 \\ 2 & 5 \end{vmatrix} + (-1) \begin{vmatrix} 1 & -1 \\ 6 & 5 \end{vmatrix} - (-3) \begin{vmatrix} 1 & -2 \\ 6 & 2 \end{vmatrix}$

$= -3(-10 + 2) - 1(5 + 6) + 3(2 + 12)$

$= 24 - 11 + 42$

$= 55$

18. $\begin{vmatrix} 5 & 9 & 4 \\ 5 & 0 & 0 \\ -6 & 4 & 1 \end{vmatrix} = -5 \begin{vmatrix} 9 & 4 \\ 4 & 1 \end{vmatrix} + 0 \begin{vmatrix} 5 & 4 \\ -6 & 1 \end{vmatrix} - 0 \begin{vmatrix} 5 & 9 \\ -6 & 4 \end{vmatrix}$

$= -5(9 - 16) + 0(5 + 24) - 0(20 + 54)$

$= 35 + 0 - 0$

$= 35$

20. $\begin{vmatrix} -3 & 2 & 1 \\ 2 & -4 & 0 \\ -1 & 8 & 1 \end{vmatrix} = 1 \begin{vmatrix} 2 & -4 \\ -1 & 8 \end{vmatrix} - 0 \begin{vmatrix} -3 & 2 \\ -1 & 8 \end{vmatrix} + 1 \begin{vmatrix} -3 & 2 \\ 2 & -4 \end{vmatrix}$

$= 1(16 - 4) - 0(-24 + 2) + 1(12 - 4)$

$= 12 - 0 + 8$

$= 20$

22. $\begin{vmatrix} -1 & -3 & 4 \\ -2 & -1 & 3 \\ 1 & 2 & 1 \end{vmatrix} = 1 \begin{vmatrix} -3 & 4 \\ -1 & 3 \end{vmatrix} - 2 \begin{vmatrix} -1 & 4 \\ -2 & 3 \end{vmatrix} + 1 \begin{vmatrix} -1 & -3 \\ -2 & -1 \end{vmatrix}$

$= 1(-9 + 4) - 2(-3 + 8) + 1(1 - 6)$

$= -5 - 10 - 5$

$= -20$

24. $\begin{vmatrix} 4 & -7 & 1 \\ 0 & 0 & 0 \\ 3 & -2 & 6 \end{vmatrix} = -0 \begin{vmatrix} -7 & 1 \\ -2 & 6 \end{vmatrix} + 0 \begin{vmatrix} 4 & 1 \\ 3 & 6 \end{vmatrix} - 0 \begin{vmatrix} 4 & -7 \\ 3 & -2 \end{vmatrix}$

$= -0 + 0 - 0$

$= 0$

26. $\begin{vmatrix} 1 & 3 & 3 \\ 2 & 6 & 4 \\ -1 & -3 & -2 \end{vmatrix} = 1 \begin{vmatrix} 6 & 4 \\ -3 & -2 \end{vmatrix} - 3 \begin{vmatrix} 2 & 4 \\ -1 & -2 \end{vmatrix} + 3 \begin{vmatrix} 2 & 6 \\ -1 & -3 \end{vmatrix}$

$= 1(-12 + 12) - 3(-4 + 4) + 3(-6 + 6)$

$= 0 - 0 + 0$

$= 0$

28. $\begin{vmatrix} 1 & 0 & 1 \\ a & b & c \\ 2 & 2 & 2 \end{vmatrix} = 1 \begin{vmatrix} b & c \\ 2 & 2 \end{vmatrix} - 0 \begin{vmatrix} a & c \\ 2 & 2 \end{vmatrix} + 1 \begin{vmatrix} a & b \\ 2 & 2 \end{vmatrix}$

$\quad\quad\quad = 1(2b - 2c) - 0 + 1(2a - 2b)$

$\quad\quad\quad = 2a - 2c$

30. $\begin{vmatrix} 1 & -3 & 4 & 2 \\ 0 & 2 & 0 & -1 \\ -2 & 0 & -1 & 2 \\ 4 & 3 & 3 & 1 \end{vmatrix}$

$= - 0 \begin{vmatrix} -3 & 4 & 2 \\ 0 & -1 & 2 \\ 3 & 3 & 1 \end{vmatrix} + 2 \begin{vmatrix} 1 & 4 & 2 \\ -2 & -1 & 2 \\ 4 & 3 & 1 \end{vmatrix} - 0 \begin{vmatrix} 1 & -3 & 2 \\ -2 & 0 & 2 \\ 4 & 3 & 1 \end{vmatrix} - 1 \begin{vmatrix} 1 & -3 & 4 \\ -2 & 0 & -1 \\ 4 & 3 & 3 \end{vmatrix}$

$= 2 \begin{vmatrix} 1 & 4 & 2 \\ -2 & -1 & 2 \\ 4 & 3 & 1 \end{vmatrix} - \begin{vmatrix} 1 & -3 & 4 \\ -2 & 0 & -1 \\ 4 & 3 & 3 \end{vmatrix}$

$= 2[1 \begin{vmatrix} -1 & 2 \\ 3 & 1 \end{vmatrix} + 2 \begin{vmatrix} 4 & 2 \\ 3 & 1 \end{vmatrix} + 4 \begin{vmatrix} 4 & 2 \\ -1 & 2 \end{vmatrix}] - [2 \begin{vmatrix} -3 & 4 \\ 3 & 3 \end{vmatrix} + 1 \begin{vmatrix} 1 & -3 \\ 4 & 3 \end{vmatrix}]$

$= 2[- 7 - 4 + 40] - [- 42 + 15]$

$= 2[29] - [- 27]$

$= 85$

32. $\begin{vmatrix} 0 & b_1 & c_1 \\ 0 & b_2 & c_2 \\ 0 & b_3 & c_3 \end{vmatrix} = 0 \cdot \begin{vmatrix} b_2 & c_2 \\ b_3 & c_3 \end{vmatrix} - 0 \cdot \begin{vmatrix} b_1 & c_1 \\ b_3 & c_3 \end{vmatrix} + 0 \cdot \begin{vmatrix} b_1 & c_1 \\ b_2 & c_2 \end{vmatrix}$

$\quad\quad\quad\quad\quad = 0 - 0 + 0$

$\quad\quad\quad\quad\quad = 0$

Determinant is zero in each case.

34. $\begin{vmatrix} ka_1 & b_1 \\ ka_2 & b_2 \end{vmatrix} = ka_1b_2 - ka_2b_1 = k(a_1b_2 - a_2b_1) = k \begin{vmatrix} a_1 & b_1 \\ a_2 & b_2 \end{vmatrix}$

and

$\begin{vmatrix} a_1 & kb_1 \\ a_2 & kb_2 \end{vmatrix} = a_1kb_2 - a_2kb_1 = k(a_1b_2 - a_2b_1) = k \begin{vmatrix} a_1 & b_1 \\ a_2 & b_2 \end{vmatrix}$.

36. $\begin{vmatrix} a_2 & b_2 \\ a_1 & b_1 \end{vmatrix} = a_2b_1 - a_1b_2 = -(a_1b_2 - a_2b_1) = - \begin{vmatrix} a_1 & b_1 \\ a_2 & b_2 \end{vmatrix}$.

38. $\begin{vmatrix} a & c \\ ka + b & kc + d \end{vmatrix} = a(kc + d) - (ka + b)c = ad - bc = \begin{vmatrix} a & c \\ b & d \end{vmatrix}$.

If one row of a second-order determinant is multiplied by a constant and added to the other row, the value of the determinant remains unchanged.

40. $\begin{vmatrix} x & y \\ y & x \end{vmatrix} = 1$

$x^2 - y^2 = 1$

Section 10.4

2. $D = \begin{vmatrix} 3 & 2 \\ 1 & 4 \end{vmatrix} = 10$

$D_x = \begin{vmatrix} 8 & 2 \\ 6 & 4 \end{vmatrix} = 20$

$D_y = \begin{vmatrix} 3 & 8 \\ 1 & 6 \end{vmatrix} = 10$

$x = \dfrac{D_x}{D} = 2$

$y = \dfrac{D_y}{D} = 1$

4. $D = \begin{vmatrix} 4 & 1 \\ 1 & 2 \end{vmatrix} = 7$

$D_x = \begin{vmatrix} 1 & 1 \\ -5 & 2 \end{vmatrix} = 7$

$D_y = \begin{vmatrix} 4 & 1 \\ 1 & -5 \end{vmatrix} = -21$

$x = \dfrac{D_x}{D} = 1$

$y = \dfrac{D_y}{D} = -3$

6.
$$D = \begin{vmatrix} 3 & -2 \\ 6 & 1 \end{vmatrix} = 15$$

$$D_x = \begin{vmatrix} 1 & -2 \\ 2 & 1 \end{vmatrix} = 5$$

$$D_y = \begin{vmatrix} 3 & 1 \\ 6 & 2 \end{vmatrix} = 0$$

$$x = \frac{1}{3}$$

$$y = 0$$

8.
$$D = \begin{vmatrix} 1 & -1 \\ 3 & 3 \end{vmatrix} = 6$$

$$D_x = \begin{vmatrix} -1 & -1 \\ 5 & 3 \end{vmatrix} = 2$$

$$D_y = \begin{vmatrix} 1 & -1 \\ 3 & 5 \end{vmatrix} = 8$$

$$y = \frac{1}{3}$$

$$y = \frac{4}{3}$$

10.
$$D = \begin{vmatrix} -2 & 3 \\ 4 & -6 \end{vmatrix} = 0$$

Since D = 0, Cramer's rule does not apply.

12. $\frac{1}{2}x + \frac{1}{3}y = -2 \Rightarrow 3x + 2y = -12$

$\frac{1}{4}x - \frac{1}{6}y = 1 \Rightarrow 3x - 2y = 12$

$$D = \begin{vmatrix} 3 & 2 \\ 3 & -2 \end{vmatrix} = -12$$

$$D_x = \begin{vmatrix} -12 & 2 \\ 12 & -2 \end{vmatrix} = 0$$

$$D_y = \begin{vmatrix} 3 & -12 \\ 3 & 12 \end{vmatrix} = 72$$

$$x = 0$$

$$y = -6$$

14.

$$D = \begin{vmatrix} 1 & 1 & 1 \\ 1 & -1 & -1 \\ 1 & 1 & -1 \end{vmatrix} = 4$$

$$D_x = \begin{vmatrix} 6 & 1 & 1 \\ 0 & -1 & -1 \\ 4 & 1 & -1 \end{vmatrix} = 12$$

$$D_y = \begin{vmatrix} 1 & 6 & 1 \\ 1 & 0 & -1 \\ 1 & 4 & -1 \end{vmatrix} = 8$$

$$D_z = \begin{vmatrix} 1 & 1 & 6 \\ 1 & -1 & 0 \\ 1 & 1 & 4 \end{vmatrix} = 4$$

$$x = \frac{D_x}{D} = 3$$

$$y = \frac{D_y}{D} = 2$$

$$z = \frac{D_z}{D} = 1$$

16.

$$D = \begin{vmatrix} 5 & 1 & 1 \\ -1 & -1 & -2 \\ 3 & 1 & 3 \end{vmatrix} = -6$$

$$D_x = \begin{vmatrix} 0 & 0 & 0 \\ -4 & -1 & -2 \\ 2 & 1 & 3 \end{vmatrix} = 6$$

$$D_y = \begin{vmatrix} 5 & 0 & 1 \\ -1 & -4 & -2 \\ 3 & 2 & 3 \end{vmatrix} = -30$$

$$D_z = \begin{vmatrix} 5 & 1 & 0 \\ -1 & -1 & -4 \\ 3 & 1 & 2 \end{vmatrix} = 0$$

$$x = \frac{D_x}{D} = -1$$

$$y = \frac{D_y}{D} = 5$$

$$z = \frac{D_z}{D} = 0$$

18.

$$D = \begin{vmatrix} 2 & -1 & 3 \\ 2 & -1 & 1 \\ 4 & 3 & -1 \end{vmatrix} = 20$$

$$D_x = \begin{vmatrix} 4 & -1 & 3 \\ 0 & -1 & 1 \\ -1 & 3 & -1 \end{vmatrix} = -10$$

$$D_y = \begin{vmatrix} 2 & 4 & 3 \\ 2 & 0 & 1 \\ 4 & -1 & -1 \end{vmatrix} = 20$$

$$D_z = \begin{vmatrix} 2 & -1 & 4 \\ 2 & -1 & 0 \\ 4 & 3 & -1 \end{vmatrix} = 40$$

$$x = -\frac{1}{2}$$

$$y = 1$$

$$z = 2$$

20.

$$D = \begin{vmatrix} 1 & 1 & 0 \\ 1 & 0 & -1 \\ 0 & 4 & 1 \end{vmatrix} = 3$$

$$D_x = \begin{vmatrix} 2 & 1 & 0 \\ 5 & 0 & -1 \\ 0 & 4 & 1 \end{vmatrix} = 3$$

$$D_y = \begin{vmatrix} 1 & 2 & 0 \\ 1 & 5 & -1 \\ 0 & 0 & 1 \end{vmatrix} = 3$$

$$D_z = \begin{vmatrix} 1 & 1 & 2 \\ 1 & 0 & 5 \\ 0 & 4 & 0 \end{vmatrix} = -12$$

$$x = 1$$

$$y = 1$$

$$z = -4$$

Section 10.5

2. $x^2 + y^2 = 25$
   $3x - y = 10$

$$-y = 10 - 3x$$

$$y = 3x - 10$$

$$x^2 + (3x - 10)^2 = 25$$

$$x^2 + 9x^2 - 60x + 100 = 25$$

$$10x^2 - 60x + 75 = 0$$

$$2x^2 - 12x + 15 = 0$$

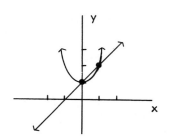

$$x = \frac{12 \pm \sqrt{144 - 4(2)(15)}}{2(2)}$$

$$x = \frac{12 \pm \sqrt{24}}{4}$$

$$x = \frac{12 \pm 2\sqrt{6}}{4}$$

$$x = \frac{6 + \sqrt{6}}{2} \qquad\qquad x = \frac{6 - \sqrt{6}}{2}$$

$$y = 3\left(\frac{6 + \sqrt{6}}{2}\right) - 10 \qquad y = 3\left(\frac{6 - \sqrt{6}}{2}\right) - 10$$

$$= \frac{-2 + 3\sqrt{6}}{2} \qquad\qquad = \frac{-2 - 3\sqrt{6}}{2}$$

4. $y = x^2 + 1$
   $y = x + 1$

$$x^2 + 1 = x + 1$$

$$x^2 - x = 0$$

$$x(x - 1) = 0$$

$$x = 0 \qquad x = 1$$

$$y = 1 \qquad y = 2$$

6. $xy = 4$
     $y - x = 0$

     $y = x$

     $xx = 4$

     $x^2 = 4$

     $x = 2 \qquad x = -2$

     $y = 2 \qquad y = -2$

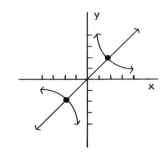

8. $y = x^2 + 2x + 1$
          $y = 2x$

$x^2 + 2x + 1 = 2x$

$x^2 + 1 = 0$

$x^2 = -1$

$x = \pm \sqrt{-1}$

$x = i \qquad x = -i$

$y = 2i \qquad y = -2i$

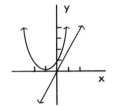

10. $9x^2 + 4y^2 = 36$
     $x^2 + y^2 = 9$

$9x^2 + 4y^2 = 36$

$\underline{-4x^2 - 4y^2 = -36}$

$5x^2 \qquad = 0$

$x^2 = 0$

$x = 0$

$0^2 + y^2 = 9$

$y = \pm 3$

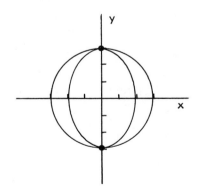

12. $x^2 + y^2 = 16$
$\underline{4x^2 - y^2 = 4}$
$5x^2 = 20$

$x^2 = 4$

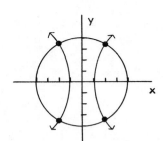

$x = 2$ $\qquad\qquad$ $x = -2$

$x^2 + y^2 = 16$ $\qquad$ $(-2)^2 + y^2 = 16$

$y^2 = 12$ $\qquad\qquad$ $y^2 = 12$

$y = \pm \sqrt{12}$ $\qquad\qquad$ $y = \pm \sqrt{12}$

$= \pm 2\sqrt{3}$ $\qquad\qquad$ $= \pm 2\sqrt{3}$

14. $xy = 6$
$x^2 + y^2 = 13$

$y = \dfrac{6}{x}$

$x^2 + \left(\dfrac{6}{x}\right)^2 = 13$

$x^2 + \dfrac{36}{x^2} = 13$

$x^4 + 36 = 13x^2$

$x^4 - 13x^2 + 36 = 0$

$(x^2 - 4)(x^2 - 9) = 0$

$x^2 = 4 \qquad\qquad x^2 = 9$

$x = 2 \qquad x = -2 \qquad x = 3 \qquad x = -3$

$y = 3 \qquad y = -3 \qquad y = 2 \qquad y = -2$

16. $x^2 + 4y^2 = 32$
    $x + 2y = 4$

$$x = 4 - 2y$$

$$(4 - 2y)^2 + 4y^2 = 32$$

$$16 - 16y + 4y^2 + 4y^2 = 32$$

$$8y^2 - 16y - 16 = 0$$

$$y^2 - 2y - 2 = 0$$

$$y = \frac{2 \pm \sqrt{4 - 4(1)(-2)}}{2(1)}$$

$$y = \frac{2 \pm \sqrt{12}}{2}$$

$$y = \frac{2 \pm 2\sqrt{3}}{2}$$

$$y = 1 + \sqrt{3} \qquad\qquad y = 1 - \sqrt{3}$$

$$x = 4 - 2(1 + \sqrt{3}) \qquad x = 4 - 2(1 - \sqrt{3})$$

$$= 2 - 2\sqrt{3} \qquad\qquad = 2 + 2\sqrt{3}$$

18. $3x^2 + 9y^2 = 13$
    $3y + 2x^2 = 7$

$$6x^2 + 18y^2 \qquad = 26$$
$$\underline{-6x^2 \qquad\qquad - 9y = -21}$$
$$18y^2 - 9y = 5$$

$$18y^2 - 9y - 5 = 0$$

$$(3y + 1)(6y - 5) = 0$$

$$y = -\frac{1}{3} \qquad\qquad y = \frac{5}{6}$$

$$3\left(-\frac{1}{3}\right) + 2x^2 = 7 \qquad 3\left(\frac{5}{6}\right) + 2x^2 = 7$$
$$2x^2 = 8 \qquad\qquad 2x^2 = 9/2$$
$$x^2 = 4 \qquad\qquad x^2 = 9/4$$
$$x = \pm 2 \qquad\qquad x = \pm 3/2$$

20. $x^2 + 2xy - y^2 = 7$

$\qquad x^2 - y^2 = 3$

$x^2 + 2xy - y^2 = 7$

$\underline{-x^2 \qquad + y^2 = -3}$

$\qquad 2xy \qquad = 4$

$xy = 2$

$y = \dfrac{2}{x}$

$x^2 - (\dfrac{2}{x})^2 = 3$

$x^2 - \dfrac{4}{x^2} = 3$

$x^4 - 4 = 3x^2$

$x^4 - 3x^2 - 4 = 0$

$(x^2 + 1)(x^2 - 4) = 0$

$x^2 = -1 \qquad\qquad x^2 = 4$

$x = i \qquad x = -i \qquad x = 2 \qquad x = -2$

$y = \dfrac{2}{i} \qquad y = \dfrac{2}{-i} \qquad y = 1 \qquad y = -1$

$= -2i \qquad = 2i$

22.

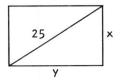

$$x = \text{width}$$

$$y = \text{length}$$

$$xy = 12$$

$$x^2 + y^2 = 25$$

$$y = \frac{12}{x}$$

$$x^2 + (\frac{12}{x})^2 = 25$$

$$x^2 + \frac{144}{x^2} = 25$$

$$x^4 + 144 = 25x^2$$

$$x^4 - 25x^2 + 144 = 0$$

$$(x^2 - 9)(x^2 - 16) = 0$$

$$x^2 = 9 \qquad\qquad x^2 = 16$$

x = 3	x = -3	x = 4	x = -4
y = 4	y = -4	y = 3	y = -3

The width is 3m and the length is 4m.

24. The simultaneous system

$$y = x + b$$
$$x^2 + y^2 = 2$$

must have only one solution.

$$x^2 + (x + b)^2 = 2$$

$$x^2 + x^2 + 2bx + b^2 = 2$$

$$2x^2 + 2bx + b^2 - 2 = 0$$

Set discriminant equal to zero.

$$4b^2 - 4(2)(b^2 - 2) = 0$$

$$4b^2 - 8b^2 + 16 = 0$$

$$-4b^2 = -16$$

$$b^2 = 4$$

$$b = \pm\, 2$$

Section 10.6

2.

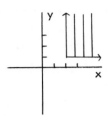

4.

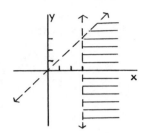

6.

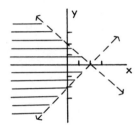

8.

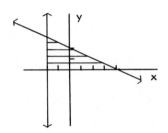

10.

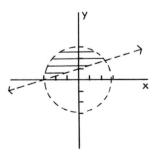

12.

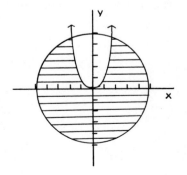

14.

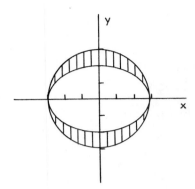

16.

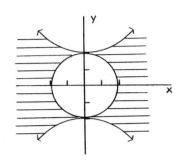

18.

20. $x + y \leq 300$
$x \geq 2y$
$x \geq 0$
$y \geq 0$

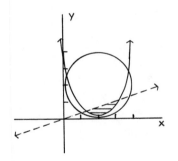

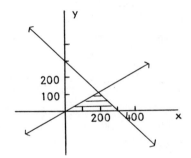

Section 11.1

2. 1, 3, 5, 7, 9

4. $2, \frac{3}{2}, \frac{4}{3}, \frac{5}{4}, \frac{6}{5}$

6. 0, 1, 3, 6, 10

8. 1, -1, 1, -1, 1

10. $-1, \frac{1}{4}, -\frac{1}{9}, \frac{1}{16}, -\frac{1}{25}$

12. $a_{25} = \frac{25(25 - 1)}{2} = 300$

14. $a_{10} = \frac{(-1)^{10}}{10^2} = \frac{1}{100}$

16. $a_n = 2n - 1$

18. $a_n = n^2$

20. $a_n = \frac{n}{n + 1}$

22. $a_n = (-1)^{n+1} n x^n$

24. $\sum\limits_{k=1}^{4} 3k = 3 + 6 + 9 + 12 = 30$

26. $\sum\limits_{k=1}^{5} (k^2 - 1) = 0 + 3 + 8 + 15 + 24 = 50$

28. $\sum\limits_{k=1}^{4} k(k - 1) = 0 + 2 + 6 + 12 = 20$

30. $\sum\limits_{k=1}^{3} \frac{(-1)^k}{k} = (-1) + \frac{1}{2} + (-\frac{1}{3}) = -\frac{5}{6}$

32. $\sum\limits_{k=1}^{3} \frac{(-1)^k}{k + 1} = (-\frac{1}{2}) + \frac{1}{3} + (-\frac{1}{4}) = -\frac{5}{12}$

34. $\sum\limits_{k=1}^{6} 2^k$

36. $\sum\limits_{k=1}^{4} \frac{1}{k}$

38. $\sum\limits_{k=1}^{5} \frac{2k - 1}{3^k}$

40. $\sum\limits_{k=1}^{5} \frac{(-1)^{k+1} x^{2k}}{2k}$

42. $a_1 = 3$

$a_n = 2a_{n-1}$

$a_2 = 2a_{2-1} = 2a_1 = 6$

$a_3 = 2a_{3-1} = 2a_2 = 12$

$a_4 = 2a_{4-1} = 2a_3 = 24$

$a_5 = 2a_{5-1} = 2a_4 = 48$

44. $\displaystyle\sum_{k=1}^{n} (a_k + b_k) = (a_1 + b_1) + (a_2 + b_2) + \cdots + (a_n + b_n)$

$$= (a_1 + a_2 + \cdots + a_n) + (b_1 + b_2 + \cdots + b_n)$$

$$= \sum_{k=1}^{n} a_k + \sum_{k=1}^{n} b_k$$

46. $\displaystyle\sum_{i=1}^{10} \frac{x_i}{n} = x_1 + \frac{x_2}{2} + \frac{x_3}{3} + \frac{x_4}{4} + \frac{x_5}{5} + \frac{x_6}{6} + \frac{x_7}{7} + \frac{x_8}{8} + \frac{x_9}{9} + \frac{x_{10}}{10}$

Section 11.2

2. $d = 2$; 11, 13, 15

4. $d = \frac{1}{3}$; $\frac{8}{3}$, 3, $\frac{10}{3}$

6. Not arithmetic

8. $d = 4$; 7, 11, 15

10. $d = -\frac{1}{2}$; $x - 2$, $x - \frac{5}{2}$, $x - 3$

12. $a_n = a_1 + (n - 1)d$

$a_n = 2 + (n - 1)5$

$a_n = 5n - 3$

14. $a_n = a_1 + (n - 1)d$

$a_n = \frac{1}{3} + (n - 1)(-\frac{2}{3})$

$a_n = 1 - \frac{2}{3}n$

16. $a_n = a_1 + (n - 1)d$

$a_n = 2x + (n - 1)1$

$a_n = 2x + n - 1$

18. $a_n = a_1 + (n - 1)d$

$a_{11} = 8 + (11 - 1)7$

$= 78$

20. $a_n = a_1 + (n - 1)d$

$a_{25} = 15 + (25 - 1)(-4)$

$= -81$

22. $a_n = a_1 + (n - 1)d$

$a_{101} = \frac{1}{5} + (101 - 1)\frac{8}{5}$

$= \frac{801}{5}$

24. $S_n = \frac{n}{2}[2a_1 + (n - 1)d]$

$S_{10} = \frac{10}{2}[2 \cdot 5 + (10 - 1)6]$

$= 320$

26. $S_n = \frac{n}{2}[2a_1 + (n - 1)d]$

$S_{10} = \frac{10}{2}[2 \cdot \frac{1}{2} + (10 - 1)\frac{1}{3}]$

$= 20$

28. $a_n = 2n + 4$

$a_1 = 6$

$a_{10} = 24$

$S_n = \frac{n}{2}(a_1 + a_n)$

$S_{10} = \frac{10}{2}(6 + 24)$

$= 150$

30. $a_n = a_1 + (n - 1)d$

$99 = 1 + (n - 1)2$

$98 = (n - 1)2$

$49 = n - 1$

$n = 50$

32. $a_n = a_1 + (n - 1)d$

$58 = 6 + (n - 1)4$

$52 = (n - 1)4$

$13 = n - 1$

$n = 14$

34. $a_n = a_1 + (n - 1)d$

$\frac{73}{4} = \frac{1}{4} + (n - 1)\frac{3}{4}$

$73 = 1 + (n - 1)3$

$72 = (n - 1)3$

$24 = n - 1$

$n = 25$

36. $\sum\limits_{k=1}^{100} (3k - 1) = 2 + 5 + 8 + \cdots + 299$

$S_n = \frac{n}{2}(a_1 + a_n)$

$S_{100} = \frac{100}{2}(2 + 299) = 15,050$

38. $11 + 22 + 33 + \cdots + 198$

$a_n = a_1 + (n - 1)d$

$198 = 11 + (n - 1)11$

$n = 18$ (which can also be found by dividing the last term
198 by the common difference 11.)

$S_n = \frac{n}{2}(a_1 + a_n)$

$S_{18} = \frac{18}{2}(11 + 198)$

$\quad = 1881$

40. $a_n = a_1 + (n - 1)d$

$11 = a_1 + (3 - 1)d \quad \Rightarrow \quad a_1 + 2d = 11$

$32 = a_1 + (10 - 1)d \quad \Rightarrow \quad \underline{-a_1 - 9d = -32}$

$\qquad\qquad\qquad\qquad\qquad -7d = -21$

$\qquad\qquad\qquad\qquad\qquad d = 3$

$\qquad\qquad\qquad\qquad a_1 + 2(3) = 11$

$\qquad\qquad\qquad\qquad\qquad a_1 = 5$

$\qquad\qquad\qquad\qquad a_{40} = 5 + (40 - 1)3$

$\qquad\qquad\qquad\qquad\qquad = 122$

42. $a_1 = 25, d = -1$

$a_n = a_1 + (n - 1)d$

$a_{10} = 25 + (10 - 1)(-1)$

$\quad = 16$ bottles

$S_n = \frac{n}{2}(a_1 + a_n)$

$S_{25} = \frac{25}{2}(25 + 1)$

$\quad = 325$ bottles

44. $a_1 = 275, d = 15$

$a_n = a_1 + (n - 1)d$

$a_{10} = 275 + (10 - 1)15$

$\quad = 410$ units

$S_n = \frac{4}{2}(a_1 + a_n)$

$S_{10} = \frac{10}{2}(275 + 410)$

$\quad = 3425$ units

46. $a_1 = 100, d = 50$

$$S_n = \frac{n}{2}[2a_1 + (n-1)d]$$

$$S_7 = \frac{7}{2}[2\cdot100 + (7-1)50]$$

$$= \$1750$$

48. $a_1 = 400, d = 50$

$$a_n = a_1 + (n-1)d$$

$$a_{13} = 400 + (13-1)50$$

$$= \$1000 \text{ per mo.}$$

50. $S = \sum_{k=1}^{n} (2k-1) = 1 + 3 + 5 + \cdots + (2n-1)$

$$= \frac{n}{2}[a_1 + a_n]$$

$$= \frac{n}{2}[1 + (2n-1)]$$

$$= \frac{n}{2}[2n]$$

$$= n^2.$$

Section 11.3

2. Not geometric

4. Not geometric

6. $r = 3$; 243

8. $r = -\frac{1}{2}$; $-\frac{1}{16}$

10. $r = \frac{2}{x}$; $\frac{8}{x^2}$

12. $a_n = a_1 r^{n-1}$

$a_n = (-1)\cdot 4^{n-1}$

14. $a_n = a_1 r^{n-1}$

$a_n = 1\cdot(-1)^{n-1}$

$a_n = (-1)^{n-1}$

16. $a_n = a_1 r^{n-1}$

$a_n = 1\cdot(\frac{1}{2})^{n-1}$

$a_n = (\frac{1}{2})^{n-1}$

18. $a_n = a_1 r^{n-1}$

$a_4 = 3\cdot 3^{4-1}$

$= 81$

20. $a_n = a_1 r^{n-1}$

$a_{26} = 1\cdot 1^{26-1}$

$= 1$

22. $a_n = a_1 r^{n-1}$

$a_{10} = 8 \left(\frac{1}{2}\right)^{10-1}$

$= \frac{1}{64}$

24. $S_n = \frac{a_1 - a_1 r^n}{1 - r}$

$S_6 = \frac{50 - 50\left(\frac{1}{10}\right)^6}{1 - \frac{1}{10}}$

$= \frac{555,555}{10^4}$

26. $S_n = \frac{a_1 - a_1 r^n}{1 - r}$

$S_6 = \frac{2 - 2(-3)^6}{1 - (-3)}$

$= -364$

28. $a_n = -4(2)^{n-1}$

$a_1 = -4(2)^{1-1} = -4$

$a_6 = -4(2)^{6-1} = -128$

$r = 2$

$S_n = \frac{a_1 - r a_n}{1 - r}$

$S_6 = \frac{-4 - 2(-128)}{1 - 2}$

$= -252$

30. $\displaystyle\sum_{k=1}^{10} 81\left(\frac{2}{3}\right)^k = 81\left(\frac{2}{3}\right) + 81\left(\frac{2}{3}\right)^2 + 81\left(\frac{2}{3}\right)^3 + \cdots + 81\left(\frac{2}{3}\right)^{10}$

$S_n = \frac{a_1 - r a_n}{1 - r}$

$S_{10} = \frac{81\left(\frac{2}{3}\right) - \frac{2}{3}\left[81\left(\frac{2}{3}\right)^{10}\right]}{1 - \frac{2}{3}} = \frac{116,050}{729}$

32. $a_n = a_1 r^{n-1}$

$\dfrac{2}{27} = 6(\dfrac{1}{3})^{n-1}$

$\dfrac{1}{81} = (\dfrac{1}{3})^{n-1}$

$n-1 = 4$

$n = 5$

34. $a_n = a_1 r^{n-1}$

$0.32 = 5r^{4-1}$

$0.064 = r^3$

$r = 0.4$

$a_3 = 5(0.4)^{3-1}$

$= 0.8$

36. $a_1 = 8000, r = 0.8$

$a_n = a_1 r^{n-1}$

$a_5 = 8000(0.8)^{5-1}$

$= \$3276.80$

38. $a_1 = 50{,}000, r = 1.1$

$a_n = a_1 r^{n-1}$

$a_5 = 50{,}000 (1.1)^{5-1}$

$= 73{,}205 \text{ persons}$

Section 11.4

2. $S_\infty = \dfrac{a_1}{1-r}$

$S_\infty = \dfrac{6}{1 - \dfrac{1}{3}}$

$= 9$

4. $S_\infty = \dfrac{a_1}{1-r}$

$S_\infty = \dfrac{1}{1 - \dfrac{1}{4}}$

$= \dfrac{4}{3}$

6. $S_\infty = \dfrac{a_1}{1-r}$

$S_\infty = \dfrac{2}{1 - (-\dfrac{1}{2})} = \dfrac{4}{3}$

8. No sum since $r = \dfrac{3}{2}$, which is greater than 1.

10. $S_\infty = \dfrac{a_1}{1-r}$

$S_\infty = \dfrac{\dfrac{1}{5}}{1 - (-\dfrac{3}{5})}$

$= \dfrac{1}{8}$

12. $0.22\overline{2} = 0.2 + 0.02 + 0.002 + \cdots$

$a_1 = 0.2 \quad r = 0.1$

$S_\infty = \dfrac{0.2}{1 - 0.1} = \dfrac{2}{9}$

14. $0.18\overline{18} = 0.18 + 0.0018 + 0.000018 + \cdots$

$a_1 = 0.18 \quad r = 0.01$

$S_\infty = \dfrac{0.18}{1 - 0.01} = \dfrac{2}{11}$

16. $9.\overline{9} = 9 + 0.9 + 0.09 + \cdots$

$a = 9 \quad r = 0.1$

$S_\infty = \dfrac{9}{1 - 0.1} = 10$

18. $0.\overline{135} = 0.135 + 0.000135 + \cdots$

$a_1 = 0.135 \quad r = 0.001$

$S_\infty = \dfrac{0.135}{1 - 0.001} = \dfrac{5}{37}$

20. $0.\overline{15} = 0.15 + 0.0015 + \cdots$

$a_1 = 0.15 \quad r = 0.01$

$S_\infty = \dfrac{0.15}{1 - 0.01} = \dfrac{5}{33}$

$3.\overline{15} = 3\dfrac{5}{33} = \dfrac{104}{33}$

22. $0.0\overline{6} = 0.06 + 0.006 + \cdots$

$a_1 = 0.06 \quad r = 0.1$

$S_\infty = \dfrac{0.06}{1 - 0.1} = \dfrac{1}{15}$

$0.4\overline{6} = 0.4 + 0.0\overline{6} = \dfrac{4}{10} + \dfrac{1}{15} = \dfrac{7}{15}$

24. $\displaystyle\sum_{k=1}^{\infty} 8\left(\frac{2}{3}\right)^k = 8\left(\frac{2}{3}\right) + 8\left(\frac{2}{3}\right)^2 + 8\left(\frac{2}{3}\right)^3 + \cdots$

$a_1 = 8\left(\frac{2}{3}\right),\ r = \frac{2}{3}$

$S_\infty = \dfrac{8\left(\frac{2}{3}\right)}{1 - \frac{2}{3}} = 16$

26. $6 + 6\left(\frac{3}{5}\right) + 6\left(\frac{3}{5}\right)^2 + \cdots$

$a_1 = 6 \qquad r = \frac{3}{5}$

$S_\infty = \dfrac{6}{1 - \frac{3}{5}} = 15$

Total distance $= 10 + 2S_\infty = 10 + 2(15) = 40$ ft

Section 11.5

```
              1
           1     1
         1    2    1
       1    3    3    1
     1    4    6    4    1
   1    5   10   10    5    1
```

2. $(a + b)^4 = a^4 + 4a^3b + 6a^2b^2 + 4ab^3 + b^4$

4. $(x + 3y)^3 = x^3 + 3x^2(3y) + 3x(3y)^2 + (3y)^3$

$\qquad = x^3 + 9x^2y + 27xy^2 + 27y^3$

6. $(x - 2)^5 = x^5 + 5x^4(-2) + 10x^3(-2)^3 + 10x^2(-2)^3 + 5x(-2)^4 + (-2)^5$

$\qquad = x^5 - 10x^4 + 40x^3 - 80x^2 + 80x - 32$

8. $(\frac{x}{2} - y)^4 = (\frac{x}{2})^4 + 4(\frac{x}{2})^3(-y) + 6(\frac{x}{2})^2(-y)^2 + 4(\frac{x}{2})(-y)^3 + (-y)^4$

$$= \frac{1}{16}x^4 - \frac{1}{2}x^3y + \frac{3}{2}x^2y^2 - 2xy^3 + y^4$$

10. $\binom{6}{2} = \frac{6!}{2!(6-2)!} = \frac{6 \cdot 5 \cdot 4!}{2!4!} = \frac{6 \cdot 5}{2!} = 15$

12. $\binom{6}{4} = \frac{6!}{4!(6-4)!} = \frac{6 \cdot 5 \circ 4!}{4!2!} = \frac{6 \cdot 5}{2!} = 15$

14. $\binom{9}{0} = \frac{9!}{0!(9-0)!} = \frac{9!}{0!9!} = \frac{1}{0!} = \frac{1}{1} = 1$

16. $\binom{9}{1} = \frac{9!}{1!(9-1)!} = \frac{9 \cdot 8!}{1!8!} = \frac{9}{1!} = 9$

18. $\binom{10}{7} = \frac{10!}{7!(10-7)!} = \frac{10 \cdot 9 \cdot 8 \cdot 7!}{7!3!} = \frac{10 \cdot 9 \cdot 8}{3!} = 120$

20. $\binom{20}{19} = \frac{20!}{19!(20-19)!} = \frac{20 \cdot 19!}{19!1!} = \frac{20}{1!} = 20$

22. $(x + y)^3 = \binom{3}{0}x^3 + \binom{3}{1}x^2y + \binom{3}{2}xy^2 + \binom{3}{3}y^3$

$$= x^3 + 3x^2y + 3xy^2 + y^3$$

24. $(x + y)^7 = \binom{7}{0}x^7 + \binom{7}{1}x^6y + \binom{7}{2}x^5y^2 + \binom{7}{3}x^4y^3 + \binom{7}{4}x^3y^4$

$$+ \binom{7}{5}x^2y^5 + \binom{7}{6}xy^6 + \binom{7}{7}y^7$$

$$= x^7 + 7x^6y + 21x^5y^2 + 35x^4y^3 + 35x^3y^4$$

$$+ 21x^2y^5 + 7xy^6 + y^7$$

26. $(\frac{x}{2} - 2)^6 = \binom{6}{0}(\frac{x}{2})^6 + \binom{6}{1}(\frac{x}{2})^5(-2) + \binom{6}{2}(\frac{x}{2})^4(-2)^2 + \binom{6}{3}(\frac{x}{2})^3(-2)^3$

$$+ \binom{6}{4}(\frac{x}{2})^2(-2)^4 + \binom{6}{5}(\frac{x}{2})(-2)^5 + \binom{6}{6}(-2)^6$$

$$= \frac{1}{64}x^6 - \frac{3}{8}x^5 + \frac{15}{4}x^4 - 20x^3 + 60x^2 - 96x + 64$$

28. $(x^2 - y)^4 = \binom{4}{0}(x^2)^4 + \binom{4}{1}(x^2)^3 y + \binom{4}{2}(x^2)^2 y^2 + \binom{4}{3}x^2 y^3 + \binom{4}{4}y^4$

$$= x^8 + 4x^6 y + 6x^4 y^2 + 4x^2 y^3 + y^4$$

30. $(a + b)^{11}$

$a = a \qquad b = b \qquad n = 11 \qquad k = 5 - 1 = 4$

Fifth term $= \binom{n}{k}a^{n-k}b^k$

$$= \binom{11}{4}a^7 b^4$$

$$= 330a^7 b^4$$

32. $(x - 5)^{15}$

$a = x \qquad b = -5 \qquad n = 15 \qquad k = 4 - 1 = 3$

Fourth term $= \binom{n}{k}a^{n-k}b^k$

$$= \binom{15}{3}x^{12}(-5)^3$$

$$= -56{,}875x^{12}$$

34. $(\frac{x}{2} + y)^{13}$

$a = \frac{x}{2} \qquad b = y \qquad n = 13 \qquad k = 10 - 1 = 9$

Tenth term $= \binom{n}{k}a^{n-k}b^k$

$$= \binom{13}{9}(\frac{x}{2})^4 y^9$$

$$= \frac{715}{16}x^4 y^9$$

36. $(r - 4s)^{20} = \binom{20}{0}r^{20} + \binom{20}{1}r^{19}(-4s) + \binom{20}{2}r^{18}(-4s)^2 + \cdots$

$$= r^{20} - 80r^{19}s + 3040r^{18}s^2 - \cdots$$

38. $(0.99)^{10} = (1 - 0.01)^{10}$

$\phantom{38.\ (0.99)^{10}} = \binom{10}{0}1^{10} + \binom{10}{1}1^9(-0.01) + \binom{10}{2}1^8(-0.01)^2 + \cdots$

$\phantom{38.\ (0.99)^{10}} = 1 - 0.1 + 0.0045 - \cdots$

$\phantom{38.\ (0.99)^{10}} \approx 0.904$